A Community of Practice Appro
Improving Gender Equality in Research

Bringing together the latest research among various communities of practice (disciplinary and place based as well as thematically organised), this volume reflects upon the knowledge, experience and practice gained through taking a unique community of practice approach to fostering gender equality in the sectors of research and innovation, and higher education in Europe and beyond. Based on the research funded by the European Union, it considers how inter-organisational collaboration can foster change for gender equality through sharing of experiences of Gender Equality Plan implementation and examining the role of measures such as change-monitoring systems. As such, it will appeal to social scientists with interests in organisational change, the sociology of work and gender equality.

Rachel Palmén is Senior Researcher in the Gender and ICT Research Programme, IN3, at the Open University of Catalonia (UOC), Spain, and at Notus Barcelona, Spain. She has worked on various gender and science research projects, including GenPORT, EFFORTI, TARGET and ACT. Her main research interests include gender (in)equalities in R&I and institutional transformation. She is the coordinator of INSPIRE: The European Centre of Excellence on Inclusive Gender Equality in Research & Innovation.

Jörg Müller is Senior Researcher in the Gender and ICT Research Programme, IN3, at the Open University of Catalonia (UOC), Spain. He has coordinated several EC projects on gender equality and science, including GenPORT, GEDII and ACT. His main research interest focuses on relational and practice-based accounts of inequalities using new methodological approaches.

Routledge Research in Gender and Society

For more information about this series, please visit: https://www.routledge.com/Routledge-Research-in-Gender-and-Society/book-series/SE0271

A Community of Practice Approach to Improving Gender Equality in Research

Edited by
Rachel Palmén and Jörg Müller

Routledge
Taylor & Francis Group

LONDON AND NEW YORK

First published 2023
by Routledge
4 Park Square, Milton Park, Abingdon, Oxon OX14 4RN

and by Routledge
605 Third Avenue, New York, NY 10158

*Routledge is an imprint of the Taylor & Francis Group,
an informa business*

British Library Cataloguing-in-Publication Data
A catalogue record for this book is available from the British Library

Library of Congress Cataloging-in-Publication Data
Names: Palmen, Rachel, editor. | Müller, Jörg (Researcher in gender
    equality), editor.
Title: A community of practice approach to improving gender equality
    in research / edited by Rachel Palmén and Jörg Müller.
Description: Abingdon, Oxon ; New York, NY : Routledge, 2023. |
    Includes bibliographical references and index.
Identifiers: LCCN 2022012527 (print) | LCCN 2022012528 (ebook) |
    ISBN 9781032115658 (hardback) | ISBN 9781032126432
    (paperback) | ISBN 9781003225546 (ebook)
Subjects: LCSH: Sex discrimination in science–European Union
    countries. | Sex discrimination in higher education–European
    Union countries. | Sex discrimination against women–European
    Union countries. | Women in science–European Union countries. |
    Research–Social aspects–European Union countries. | Communities
    of practice–European Union countries.
Classification: LCC Q130 .C656 2023 (print) | LCC Q130 (ebook) |
    DDC 305.43/5094–dc23/eng20220616
LC record available at https://lccn.loc.gov/2022012527
LC ebook record available at https://lccn.loc.gov/2022012528

ISBN: 978-1-032-11565-8 (hbk)
ISBN: 978-1-032-12643-2 (pbk)
ISBN: 978-1-003-22554-6 (ebk)

DOI: 10.4324/9781003225546

Typeset in Times New Roman
by KnowledgeWorks Global Ltd.

The ACT project has received funding from the European Union's Horizon 2020 research and innovation programme under grant agreement No 788204. The contents of this book reflect only the authors' views; the Research Executive Agency (REA) and the European Commission are not responsible for any use that may be made of the information it contains.

Contents

Figures

Tables

Notes on Contributors

Amanda Aldercotte, PhD, Head of Knowledge and Research, Advance HE. Amanda completed her PhD at the University of Cambridge and has been working in higher education for the last ten years. Her most recent research explored differences in men and women's experiences of the UK higher education workplace and in the transition to home-based working following the COVID-19 pandemic.

Laufey Axelsdóttir, PhD, is a post-doctoral researcher and a sessional teacher in Gender Studies at the University of Iceland. Her main research focus is on labour market issues, such as gender quotas, recruitment policies, gender-balanced family responsibility, gender relations in organisations, and history and background of the textile industry in Iceland.

Sarah Barnard, PhD, Senior Lecturer in Sociology of Contemporary Work in the School of Business and Economics at Loughborough University, UK. Sarah has an established research profile in Equality, Diversity and Inclusion at work and interdisciplinary research on Science, Engineering and Technology contexts. Particular interests include gendered aspects of careers and work; organisational practices and policies; and the construction sector.

Sarah Beranek (F), M.Sc., has a background in Social Sciences (University of Augsburg, Germany) and Socioeconomics (WU Vienna, Austria). Since 2017, she has been working as a researcher at JOANNEUM RESEARCH in the POLICIES Institute, focusing on research on gender equality and evaluation in R&I.

Maria Caprile, M.A., is director of research at Notus, Spain. Over the last 15 years, she has directed large EU funded comparative studies on gender in research, 'Meta-analysis of gender and science research' and the FP7 project 'SHEMERA'. She is currently participating in two H2020 projects: TARGET and ACT.

Ewelina Ciaputa, MA, Researcher at the Institute of Sociology at the Jagiellonian University in Krakow, Poland. Her focus of work covers gender equality, sexuality, reproductive rights, masculinities, citizenship, and discourse. She is a co-facilitator of the Community of Practice (GEinCEE CoP) created under the framework of the ACT project.

Andrew Dainty, Professor, Pro-Vice Chancellor for Education, Manchester Metropolitan University, UK. A renowned expert on the sociologies of project-based organisations, Andy's research has focused on the social rules and processes in project teams. Andy's work is interdisciplinary with academic collaborators from across the engineering, social sciences, business, and economics fields; leading consultancy and contracting organisations, and client bodies.

Areti Damala, PhD, is an independent researcher, university lecturer, and consultant based in Paris, France. She facilitated the ACT 'STRATAGIES for Sustainable Gender Equality' Community of Practice (Strategies). She is currently (2021–2025) a member of the Management Committee of the COST Action 'Making Young Researchers' Voices Heard for Gender Equality' (CA20137).

Þorgerður J. Einarsdóttir is Professor of Gender Studies at the University of Iceland. Her research areas cover a broad range of issues from feminist theory and equality policies to labor market issues, masculinities, and gender and academia. One of her most recent projects is gender budgeting.

Anne-Sophie Godfroy, PhD, is associate professor of Philosophy and gender equality officer at University Paris-Est Créteil, and researcher at « Republique des Savoirs », a transdisciplinary joint research unit of CNRS, Ecole Normale Supérieure and Collège de France. She was the leader of the CNRS team in the ACT project. She currently chairs the COST Action VOICES (CA20137).

Kevin Guyan, University of Glasgow, UK. Kevin is a Research Fellow in the School of Culture and Creative Arts at the University of Glasgow. His work explores the intersection of data and identity, he is also the author of *Queer Data: Using Gender, Sex and Sexuality Data for Action* (Bloomsbury Academic).

Tarek M. Hassan, Professor of Construction Informatics and Associate Dean for Enterprise at the School of Architecture, Building and Civil Engineering, Loughborough University, UK. Tarek has extensive research expertise in the areas of advanced construction information technology and research into higher education with particular emphasis on gender equality.

Florian Holzinger, MA, Researcher at Joanneum Research, Policies – Institute for Economic and Innovation Research in Vienna, Austria. His research covers the management of research projects in national and international project environments, gender equality in research organisations, and female careers in research and innovation.

Lisa Kamlade works as a project manager at Deutsches Elektronen-Synchrotron (DESY) in Hamburg, Germany. She currently finishes her studies in business psychology (M.Sc.). Among other things she helps to coordinate the GENERA Network and works in the European project ACT as a Community of Practice (CoP) facilitator.

Ewa Krzaklewska, PhD, is a sociologist, and works as an Assistant Professor at the Institute of Sociology of the Jagiellonian University in Krakow, Poland. Her research interests relate to youth sociology, mobility, gender equality, and family studies. Between 2018–2021, she coordinated the Horizon 2020 ACT project at the Jagiellonian, which aimed at supporting Central and Eastern European research institutions in implementation of gender equality policies through establishing a Community of Practice @ACTonGEinCEE.

Karolina Kublickiene, MD, PhD is an Associated professor at Karolinska Institutet (KI), Stockholm, Sweden. She leads research in cardiorenal medicine towards novel treatments. She founded gendered innovation alliance at KI and has extensive portfolio about gender dimension in research and education. She lectures, and leads KI's doctoral program with state-of-the-art education. She participates in several EU projects, and frequently speaks at international events.

Jovana Mihajović Trbovc is a research fellow and gender equality advisor at the Research Centre of the Slovenian Academy of Sciences and Arts. She researches and provides trainings on the issues related to gender equality in academia, is a member of the Commission for Equal Opportunity in Science (Slovenia) and facilitator of the CoP Alt+G.

Chloé Mour is a Research Assistant at Sciences Po University (Paris). She received her Master of Sociology at the School for Advanced Studies in the Social Sciences (Paris) in 2019. In 2020, she worked as a CoP ('STRATEGIES') facilitator at the National Center for Scientific Research (Paris) within the ACT project.

Kathrin Rabsch, M.A., worked as a research associate at Technical University Berlin, Germany and is currently working at Hochschule für Technik und Wirtschaft Berlin, Germany. She has been a core partner in the ACT project. She holds a M.A. in 'Society, globalization and development' (Sociology and Political Science). She has a special interest in qualitative research, sociological theories, and social inequality research.

Sonja Reiland has a PhD in biochemistry and works as project manager at the Centre for Genomic Regulation, Barcelona, Spain. As the project manager of the H2020 project LIBRA (2015–2019), she supported 10 Life Science research centers in GEP development and implementation and in ACT (2019–2021), she facilitated the LifeSciCoP.

Sybille Reidl, M.A. is senior scientist and deputy research group leader at the Institute for Economic and Innovation Research (POLICIES) of JOANNEUM RESEARCH. Her main research areas are gender/diversity and human resources and evaluation of gender equality policies in RTDI. Moreover, she supports technology development projects in integrating gender and other diversity dimensions.

Finnborg Salome Steinþórsdóttir, PhD, is a post-doctoral researcher and sessional teacher in Gender Studies at the University of Iceland. Her main research focus is on gender budgeting and implementation of equality policies in research performing organisations. Her research interests also revolve around gender relations in organisations, organisational cultures, and gender-based violence.

Paulina Sekuła, PhD, Assistant Professor at the Institute of Sociology at the Jagiellonian University in Krakow, Poland. Her work focuses on gender inequalities in research and academia, structural change processes, and political culture. She participates in a H2020 project MINDtheGEPs aiming at reducing gender imbalances in research and academia. She is a co-facilitator of the Community of Practice (GEinCEE CoP) created under the framework of the ACT project.

Aleksandra Thomson, PhD, Research Associate in the School of Business and Economics at Loughborough University; now Research Associate in Equality, Diversity and Inclusion (EDI) at the Elizabeth Blackwell Institute for Health Research at the University of Bristol, UK. Her research interests include EDI and careers in work organisations, flexible working, work-life balance, institutional change for gender equality, and Communities of Practice.

Claartje Vinkenburg, PhD, is an independent expert consultant on (gender) diversity and inclusion. Her area of expertise is careers in research and professional service firms. Claartje is an in-house consultant with Portia and visiting fellow at Vrije Universiteit Amsterdam. Recent clients include research councils, national government, and the European Commission.

Marta Warat, PhD, assistant professor at the Institute of Sociology, Jagiellonian University in Krakow, Poland. Her main academic interests involve gender in European societies, gender equality policies, inequalities, citizenship, and democracy. Between 2018 and 2021, she was a researcher in the H2020 project ACT. She is the local coordinator in Men in Care project (EaSI Programme, EU).

Sergi Yanes Torrado, PhD. in Social Anthropology. Researcher at the Gender and ICT research group (GenTIC) of the Internet Interdisciplinary Institute (IN3) – Universitat Oberta de Catalunya (Spain), Associate Professor at the Universitat Autònoma de Barcelona (Spain) and independent consultant. Member of the Catalan Institute of Anthropology.

Acknowledgements

Firstly, we would like to thank the European Commission for providing the funding for the ACT project. This, first and foremost, enabled us to put into practice an exciting vision and realise a collaborative, innovative approach to promoting gender equality in R&I and HE. We acknowledge, and are grateful for the financial support from the Commission which made the ACT project possible as well as making this book a reality. Of course, however we take full responsibility for the views in this book.

We would also like to thank all the authors of the book for the time and effort invested into crafting the chapters for this book. This manuscript was developed during difficult times due to the global COVID-19 pandemic which resulted in some of our authors suffering illness, burn out, as well as a general zoom overload. Many of our authors were juggling full-time jobs, home schooling, and had multiple demands whilst some faced various health problems. We are grateful for the extra time and effort – that authors invested in the book, documenting what we believe has been a ground breaking approach to advancing organisational change for a greater gender equality in R&I and HE throughout Europe and Latin America.

We would also like to thank consortium members who are not authors of the book – but without their input into the ACT project – this book could not have been possible.

A special mention should also go to all the Community of Practice facilitators for their efforts in engaging with and supporting their communities throughout the duration of the project. Without their commitment, professionalism and ability to engage CoP members the project would not have been possible. We would also like to thank the more than 140 institutional and individual members of our 8 Communities of Practice for their constant engagement with the project as well as those that generously gave more of their time to be interviewed for the book.

We also thank the editing team at Routledge, Taylor and Francis Books for their professionalism and patience throughout the production process.

Abbreviations

AKKA	Gender Integrated Leadership Programme (Lund University)
Alt+G	Alternative Infrastructure for Gender Equality
AR20	Achievement Relative to Opportunity
ARRS	Scientific Council of the Slovenian Research Agency
ASSET	Athena Survey of Science, Engineering and Technology
CA	European Cooperation in Science and Technology Action
CEE	Central and Eastern Europe
CNRS	French National Centre for Scientific Research
CoP	Community of Practice
CoPs	Communities of Practice
CoPP	Community of Political Practice
CoPPs	Communities of Political Practice
COST	European Cooperation in Science and Technology
CPED	Standing Conference for Equality and Diversity
CV	Curriculum Vitae
DAKI	Drop, Add, Keep, Improve
DORA	Declaration on Research Assessment
EC	European Commission
ECRIs	Early Career Researchers and Investigators
EIGE	European Institute for Gender Equality
ERA	European Research Area
ERAC	European Research Area and Innovation Committee
ERAC SWG GRI	European Research Area and Innovation Committee Standing Working Group on Gender in Research and Innovation
ERAG	European Research Area Group - specific to the ACT project
ERC	European Research Council

EU	European Union
EUA	European Universities Association
Eurodoc	European Council of Doctoral Candidates and Junior Researchers
EURODOC	European Council of Doctoral Candidates and Junior Researchers
EWORA	European Women's Rector's Association
FF UL	Faculty of Arts, University of Ljubljana
FORGEN	Funding Organisations for Gender Equality
FP7	Seventh Framework Programme
GE	Gender Equality
GEAM	Gender Equality Audit and Monitoring
GEinCEE	Gender Equality in Central and Eastern Europe
GENERA	ACT Community of Practice in Physics
GenBUDGET	ACT Community of Practice in gender budgeting
GEP / GEPs	Gender Equality Plan(s)
HE	Higher Education
HEI	Higher Education Institution
HR	Human Resources
ICT	Information Communication Technologies
JU	Jagiellonian University
LAC	Latin American Community of Practice
LERU	League of European Research Universities
LGBT	Lesbian, Gay, Bisexual and Transgender
LifeSciCoP	ACT Community of Practice in Life Sciences
MoU	Memorandum of Understanding
MS	Member States of the European Union
NAP	National Action Plan
NIHR	National Institute for Health Research
NOGAFEM	CoP Transforming Women's Health
ONS	Office for National Statistics
PhD	Doctor of Philosophy
PoCs	Practice of Communities
R&D	Research and Development
R&I	Research and Innovation
RFOs	Research Funding Organisations
RPOs	Research Performing Organisations
RRI	Responsible Research and Innovation
RSCI	Royal College of Surgeons in Ireland
RWTH Aachen University	Rheinisch-Westfälische Technische Hochschule Aachen University
SDG	Sustainable Development Goals
SPIRIT	Slovenian Public Agency for Entrepreneurship, Internationalization, Foreign Investments and Technology

SPSS	Statistical Package for the Social Sciences
STEM	Science, Technology, Engineering and Mathematics
STEMM	Science, Technology, Engineering, Mathematics and Medicine
TIPs	Targeted Implementation Projects
UK	United Kingdom
UL	University of Łódź
USA	United States of America
WGs	Working Groups
WLB	Work Life Balance
ZRC SAZU	Research Centre of the Slovenian Academy of Sciences and Arts
ZRS	Koper Science and Research Centre Koper

1 Introduction

Reflecting on a Community of Practice approach to institutional change for a greater gender equality in R&I and HE – Policy and practice

Jörg Müller and Rachel Palmén

This book reflects on the use of Communities of Practice (CoPs) to further gender equality in research and innovation (R&I) organisations and higher education (HE) institutes throughout Europe. It is grounded on our experiences of setting up and supporting eight CoPs comprising 144 organisations as part of the ACT project (2018–2022), a three-and-a-half-year effort funded by the European Commissions' Horizon 2020 programme.[1]

The 12 chapters collected in this volume provide a window into the practical experiences and lessons learnt by CoP members, CoP facilitators and collaborators. The rich diversity of CoP organising principles (geographic, disciplinary and thematic) offers key insights into the different challenges faced by change agents in pushing for gender equality in R&I and HE. Together with a sound conceptual embedding in the CoP literature as well as the wider literature on gender equality interventions, the various perspectives presented contribute to providing a better, more nuanced understanding of the complex European landscape of gender equality in R&I and HE. A particular focus that runs throughout the book will examine how inter-organisational cooperation can be harnessed to impact the three objectives that form part of the European Research Area (ERA) priority 4 on gender equality and gender mainstreaming in R&I: scientific careers, decision-making and integrating the gender dimension in teaching and research content. These wider insights are rounded up with reflections on the benefits and limitations of a CoP approach to promoting gender equality in R&I and HE.

In 2021, it has been exactly 30 years since the idea of CoPs was introduced through the publication by Jean Lave and Etienne Wenger (1991) on "Situated Learning". A rich literature has emerged in the meantime, building upon and unfolding the three determining features of a CoP, namely, a "joint enterprise" (shared interest, or domain), "mutual engagement" (community) and development of a "shared repertoire" of resources and practice

DOI: 10.4324/9781003225546-1

(Wenger, 1998). Extensive reviews testify to the different engagements and ways of appropriating CoPs with regard to organisational embedding (Schulte, 2020), knowledge management (Bolisani & Scarso, 2014), innovation studies (Pattinson et al., 2016), social learning systems (Blackmore, 2010), higher education (McDonald & Cater-Steel, 2017) or nursing practice (Terry et al., 2020) to name just a few. Surprisingly, gender scholars are relatively absent from this body of literature. While organisational scholars have extensively contributed to the reception of this concept, experts on gender – despite their overall contribution to the literature on organisational change – have only engaged on the margins with this body of work. Except for the special edition of Language and Society (Holmes & Meyerhoff, 1999), there have been only isolated publications reflecting on gender and CoPs specifically, largely in the Women's Studies International Forum (Curnow, 2013; Paechter, 2003; Stapleton, 2001; Wagner, 1994) while mayor journal outlets such as *Gender, Work and Organisation* have remained silent on this topic. Equally, from a policy perspective, the suggestion to appropriate CoPs for gender equality work in organisations is relatively recent. While the European Commission has put great emphasis on institutional change (European Commission, 2011) highlighting Gender Equality Plans (GEPs) as the main instrument for achieving gender equality, beyond this, particular methodologies for its implementation are not defined. Explicit references to CoPs (or "communities of practitioners") only start to emerge from 2013 onwards. Notable, sporadic usage of the concept does exist with regard to the implementation of a GEP (Barnard et al., 2016); however, a more systematic and empirically grounded exploration of what CoPs are and can achieve for gender equality, particularly in the context of R&I and HE institutions, is largely missing.

Gender equality in R&I and HE: Evidence and policy framework

The present book addresses this lacuna in the literature and aims to provide an explicit consideration of CoPs as an instrument for accelerating gender equality and institutional change in R&I across Europe. So, what do we mean when we talk about gender (in)equality in R&I in Europe? Considering the statistical key figures on gender equality first, one can applaud the overall improvements over the years but lament the slow pace. According to the most recent She Figures (European Commission, 2021b), gender balance among PhD graduates (48.1% women) has nearly been reached, yet women account only for one-third of all researchers in the European Union (EU) and one-fifth in the business sector. Women are also still significantly under-represented at higher stages of the career ladder: the share of women in Grade A positions in HE (full professor and equivalent) reached just 26% for the EU in 2018 and the proportion of women heading HE institutions in Europe was only 23.6% in 2019. The number of women among patent holders also remains extremely low, similar to the low participation of women in the creation of innovative start-ups. Concerning the integration of the gender dimension, the recent data provided by She Figures are sobering: only 1.80% of scientific publications of the EU integrate a gender analysis despite the fact that an increasing number of organisations at least mention the cited actions and measures towards gender equality on their websites. Whilst gender (in) equalities in R&I and HE go far beyond a binary representation of men and women at different levels and fields of academia – the statistical evidence of the under-representation of women is one key piece of the puzzle that cannot be overlooked.

In 2012, the European Commission established gender equality as one of five priorities for achieving the objective of a common research area in Europe and this policy has been progressively strengthened (European Commission, 2020). Three objectives were established for EU countries to work on and foster institutional change:

- Gender equality in scientific careers
- Gender balance in decision-making
- Integration of the gender dimension into the content of research and innovation

It is these three gender equality objectives for institutional change that have provided a policy framework for our work in the ACT project.

The Council Conclusions on Advancing Gender Equality in the European Research Area developed in 2015 stated that EU Member States should "make institutional change a key element of their national policy framework on gender equality in R&I" by developing national action plans or strategies at both the national and institutional levels (Council of the European Union, 2015). Incentives should be provided by Member States for research-performing organisations (including universities) "to revise or develop gender mainstreaming strategies, GEPs including the gender dimension in R&I content and programmes and mobilise adequate resources to ensure their implementation". The Council Conclusions also highlight the need to strive for gender balance in leadership and decision-making positions and invite relevant authorities to establish guiding targets (i.e. quantitative objectives) to improve gender balance in decision-making bodies specifying "leading scientific and administrative boards", "recruitment and promotion committees" as well as "evaluation panels". National Action Plans were then developed in 2016 by Member States that included concrete actions to advance gender equality (Ferguson, 2021, p. 14).

The European Research Area and Innovation Committee (ERAC) Standing Working Group on Gender in Research and Innovation's (2018) main findings regarding sustainable cultural and institutional change include the following: huge differences between EU-15[2] and EU-13[3] countries, the majority of incentives tend to be introduced by national authorities and national funding agencies in the EU-15 countries and across the board whilst very few incentives have been introduced to integrate the gender dimension in research. Only four countries (all of which are in the EU-15) have introduced guiding targets for the proportion of women among professors. Concrete measures to reduce the effect of gender bias in the allocation of research funding have been put in place by only two national funding agencies (and no national authorities) in the EU-13 in contrast to six national funding agencies and three national authorities in the EU-15. Regarding gender balance in decision-making: six EU-15, two EU-13 and three associate countries have established guiding targets for gender balance, but these have not been implemented anywhere. The report not only highlights the differences between the EU-15 and the EU-13 but also notes great variation between Strong Innovators and Innovation Leaders on the one hand and Moderate and Modest Innovators on the other – to the extent to which policies and actions to advance gender equality in the ERA are implemented (or not). The high positive correlation between countries' positions on the 2018 EU Innovation Scoreboard and the 2017 Gender Equality Index is also recognised.

As the new ERA communication (European Commission, 2020) highlights, despite the robust policy framework put into place, there remain profound disparities in terms of policy implementation as well as the representation of women in R&I across Member States. As the chapters in this volume will show, CoPs can offer a new and promising bottom-up approach to complement the overarching policy frameworks with locally situated, context-dependent knowledge production and development of practical solutions.

Setting up eight Communities of Practice

The ACT project has setup and supported eight CoPs throughout its lifetime. From the very outset, the ACT Consortium was constructed to build upon the insights and networks created by various previous structural change projects funded by the European Commission (for an extensive overview of these projects see Ferguson, 2021). So-called seed partners set up one CoP, each building upon their work in structural change projects such as GARCIA, GenderTime, GENERA, INTEGER, LIBRA, SPEAR and TARGET. A CoP facilitator based within each seed partner organisation coordinated and supported the working of the CoP members through organising meetings, facilitating shared workspaces and providing the momentum for concrete equality work. Although all CoPs were constituted as a collaboration among different organisations, formalised by the signature of a Memorandum of Understanding and had a uniform governance mechanism, their thematic orientation varied considerably. As shown in Table 1.1, CoPs included different thematic foci such as gender in physics (GENERA) or the life sciences (LifeSciCoP), gender budgeting (GenBUDGET), STRATEGIES with a focus on sustainability and

Table 1.1 Overview of ACT Communities of Practice.

Acronym	Title	Focus	Coordinated by
LifeSciCoP	Gender Equality in the Life Sciences	Thematic/ disciplinary	Fundació Centre de Regulació Genòmica (CRG), Spain
GEinCEE	Gender Equality in Central and Eastern Europe	Geographic	Universytet Jagiellonski (UJ), Poland
GenBUDGET	Gender Budgeting in Research Organisations	Thematic	Haskoli Islands (UoI), Iceland
FORGEN	Funding Organisations for Gender	Thematic	Science Foundation Ireland (SFI), Ireland
GENERA	Gender Equality in Physics	Thematic/ disciplinary	Deutsches Elektronen-Synchroton (DESY) & Umweltbundesamt (UBA), Germany
STRATEGIES	Strategies for Sustainable Gender Equality	Thematic	Centre National de la Recherche Scientifique (CNRS), France
Alt+G	Alternative Infrastructure for Gender Equality	Geographic	Research Centre of the Slovenian Academy of Sciences and Arts (ZRC SAZU)
LAC	Latin American Community of Practice	Geographic	Regional UNESCO Chair-Women, Science and Technology in Latin America – FLACSO Argentina

early career researchers or for funding organisations (FORGEN). Other CoPs had a clear geographic orientation, with GEinCEE supporting gender equality work across many organisations in Eastern Europe, Alt+G in Slovenia and the Latin American Community of Practice (LAC). As a result of this diverse set of thematic, disciplinary and geographic CoPs, the insights collected in the chapters of this volume cover a lot of ground in terms of different organisational settings and challenges for gender equality in R&I throughout Europe.

To some extent, the ACT project and subsequently this book are characterised by the dual aims or logics of strengthening gender equality within R&I and HE institutions on the one hand and implementing an inter-organisational CoP approach on the other. This tension has really defined the project in terms of its overarching aim – to foster collaboration across several organisations or to really push forward gender equality within a single organisation. Whilst initially institutional change could be conceptualised as the broad goal and inter-organisational collaboration the means to achieve it, we can see how this conceptualisation may be problematic on various levels. Firstly, the whole concept of institutional change is focused on the internal workings of one organisation at a time. GEPs target individual organisation or its sub-units such as departments or faculties – each having its specific needs and idiosyncratic agendas of change. Although many examples of CoPs do exist that operate within a single organisation, the distinct approach of the ACT project consisted of setting up inter-organisational CoPs whose members span organisational boundaries. The question that naturally then arises concerns the possibilities to bring these two worlds together and apply cross-institutional learning to gender equality challenges within one's own institution. As it will become apparent, the chapters to this edited volume can be seen as contributing with their reflections to the wider discussion on an eco-system approach to foster equality and social justice beyond the individual organisation (Janssens & Zanoni, 2021).

A second issue concerns the relative autonomy of CoPs. As will become clear in the next section, CoPs are highly flexible and innovative forms of organising social learning whose effectiveness is grounded in their bottom-up, needs-centred management. CoPs are autonomous "units" which can be difficult to subsume under pre-defined, top-down organisational goals – even when these goals are as valuable as the pursuit of gender equality. A cursory reading of the CoP literature highlights the perils of superimposing objectives on CoP members that do not meet their needs. Thus, cultivating CoPs implies being attentive to the internal, bottom-up agenda setting as it unfolds in relation to overarching and broader goals such as the design and implementation of a GEP within an organisation.

What makes the chapters assembled in this book so interesting is this very tension – bringing together a gender equality lens with the CoP approach. This book offers a rich overview not only regarding the diverse CoP experiences of institutional collaboration in pushing forward the gender equality agenda but also regarding the diverse thematic issues that constitute the landscape of gender equality in R&I and HE across Europe and beyond. We think that bringing together these approaches has proven more powerful and fruitful than we could have ever predicted.

Community of Practice – Its relevance for advancing gender equality

Let's start with a minimal definition: "Communities of Practice are groups of people who share a concern, a set of problems, or a passion about a topic, and who deepen their knowledge and expertise in this area by interacting on an ongoing basis" (Wenger et al., 2002, p. 4). This definition put forward by Wenger and colleagues suggests a coherence and clarity of what CoPs entail that does not necessarily match the empirical reality which is far more complex. CoPs can differ along their lifecycle phase which run from initial design/launching to growing and maturing (Wenger et al., 2002). CoPs in the empirical field also differ in terms of their demographics (purpose and maturity), organisational context (creation process, professional/ organisational boundary-crossing and degree of institutionalisation) and membership characteristics (size, geographic dispersion and selection) to name just the most relevant features in the context of this book (Dubé et al., 2003; Hara et al., 2009). Depending on which aspect is deemed most interesting, authors have foregrounded certain features while neglecting others. Amin & Joanne (2006), for example, distinguish four types of CoPs, namely, task/craft-based CoPs which are preoccupied with the preservation of knowledge from professional- or expert/creative-based communities whose focus is on the creation of new knowledge (Pattinson et al., 2016).

As already mentioned, the eight CoPs in the ACT project are indeed relatively uniform: all of them are inter-organisational CoPs involving persons that are located across different organisations. This also implies that the CoPs are relatively dispersed geographically speaking: while in the case of Alt+G membership spans several organisations in the same country (Slovenia), in other cases members are distributed across a certain geographic region like several Eastern European (GEinCEE) or mainly Northern European (GenBudget) countries, or the entire South American continent (LAC). Although several CoPs are a prolongation of previous structural change projects, none of them has been launched as a CoP for longer than three years. This implies that all CoPs within ACT pertain to an early lifecycle phase, with a rather limited lifespan due to the end of the project funding in 2022. In all cases, members within the ACT CoPs are quite diverse, usually spanning organisational, disciplinary and cultural backgrounds – which provide a rich and diverse environment for mutual learning.

These empirical features of the ACT CoPs need to be put in dialogue with the conceptual dimensions and issues discussed in the wider literature. By carrying out a selective reading of the three foundational facets of CoPs – domain, community and practice – in conjunction with the gender equality literature, the starting points for conceiving CoPs as an instrument for advancing gender equality in R&I and HE in Europe and beyond will become into sharper view.

The domain: Knowledge and gender equality

As already mentioned, a CoP is defined first, through a "domain" or shared interest among its participants. This domain of knowledge "creates a common ground and a sense of common identity" and "inspires members to contribute and participate, guides their learning, and gives meaning to their actions" (Wenger et al., 2002, p. 27). Rather than simply being a stated goal, what differentiates a CoP from a project team, for example, are relations of "mutual accountability" towards its subject domain. It implies generating knowledge through nurturing and re-negotiating a shared understanding of what is important.

As a consequence, what a CoP "is" or "can do" is fundamentally entangled with how knowledge and learning are conceived. Indeed, the revolutionary impulse from the outset of Lave and Wenger's book on "peripheral participation" (1991) consisted of moving beyond a cognitive account of learning towards a social process-based model. Learning, in this initial account, was not conceived as a mental exercise of appropriating explicit, codified knowledge but rather as a gradual transition from "peripheral to full membership" in a (professional) community. While learning through social participation can involve episodes of transmission of codified knowledge (facts, theories), it also and more importantly involves apprenticeship through supervised, hands-on practice. Knowledge is never simply transferred from expert to novice but requires interactions among "oldtimers" and "newcomers" involving the observation of codes of conduct as well as the imitation of how things are done. In short, it requires a whole set of practices that need to be learned through (social) interaction and participation.

The emphasis on this social dimension of knowledge strikes an immediate chord with feminist thinking. Different philosophers of science have argued that knowledge is socially situated (Anderson, 1995; Harding, 1986; Longino, 1990). However, instead of underscoring simply the social embeddedness of learning, gender scholars have highlighted the resulting partial and biased nature of knowledge, foregrounding ultimately the political dimension of all knowledge claims (Haraway, 1988). As Alison Wylie writes, "social location systematically shapes and limits what we know, including tacit, experiential knowledge as well as explicit understanding, what we take knowledge to be as well as specific epistemic content" (Wylie, 2003, p. 31). First formulated during the 1970s and 1980s and refined through contemporary debates, standpoint feminism (Harding, 2004; Intemann, 2010) leaves no doubt that social positions in society are hierarchically structured by power relations which in turn condition not only individual experiences but also the means to make collective sense of these. Knowledge, far from being a neutral and distanced accumulation of facts and universal laws, involves political negotiations of value and struggles over what is included or excluded, what/who is in positions of power and what/who is operating on the margins of science and society.

Both aspects – the standpoint dependent production of knowledge as well as its concomitant political and power dimension – have been discussed in the CoP literature, albeit to different degrees. The insight into the situatedness of learning and knowledge is tightly associated with the concept of practice – which always conceives social interactions as embedded in a network of material artefacts and objects (see also section on practice below). The fact that learning is always located in an idiosyncratic social context constitutes a prominent point of departure for early receptions of Lave and Wenger's (1991) work. Thus, Brown and Duguid (1991), for example, underscore the strength of CoPs in being responsive and flexible to address unforeseen and emergent challenges in work practice. By conceiving learning as rooted in social practice, CoPs become a highly effective, organic instrument of innovation as practical solutions are generated where they emerge, continuously refined in tight, localised feedback loops until the job is done. Undoubtedly, the ability to take one's immediate needs and interests as a starting point for CoPs was one of the key motivating factors to invest in this type of work among the participants of the ACT CoPs.

However, the idiosyncratic nature of CoPs – their responsiveness to local context – posed from the very outset also a key challenge particularly for management scholars in terms of steering and controlling the ensuing innovation process. How can locally generated solutions to problems be re-inserted and aligned with the overarching organisational goals? Unfortunately, as Schulte (2020) and others have remarked, the implied power relations both within CoPs as well as in relation to their wider organisational, political, legal and cultural embedding have so far not been sufficiently addressed (Contu & Willmott, 2003; Schulte, 2020). The fact that many CoPs emerge through bottom-up processes does not imply that they exist in a power-free vacuum, neither among its members nor in terms of their knowledge production. That knowledge is contested should be nowhere more visible than in the arena of gender equality and institutional change. On the one hand, as gender equality work often involves academic as well as administrative staff across organisational as well as scientific units, what counts as knowledge and evidence for decision-making is not self-evident. In addition, CoPs tend to operate outside the formal, established organisational units – which is an advantage when dealing with a transversal issue such as gender. However, insofar gender equality work aims for a redistribution of resources and privileges – it is also likely to come into direct conflict with wider organisational agendas, goals and decision-making power. The well-rehearsed insistence to include top management and decision-makers in gender equality work points in this direction, to assure CoPs leverage in terms of organisational steering and decision-making. Producing knowledge through CoPs is insufficient without the ability to make decisions based upon this knowledge for greater gender equality.

The reflection of the situated and political knowledge creation with/ through CoPs also needs to be critically examined from a European policy-level perspective. The experiences and knowledge that will emerge across CoPs that operate in different national contexts bring into sharper focus what can be learned across these national contexts and across CoP experiences. Wenger-Trayner and colleagues introduce the concept of "Landscape of Practice" to explain how different CoPs might interact and depend upon each other rather than their own, situated practices (Pyrko et al., 2019; Wenger-Trayner et al., 2014). However, this conjures up the question of which knowledge is considered "valid" knowledge? What might be deemed important in one context does not necessarily apply in another one. Hence, the simple generation of knowledge within and across CoPs becomes a political negotiation about the empirical adequacy of what is important, what counts and what serves as evidence for subsequent actions and policies. What can we learn from the situated knowledge generated in Sweden for our situation in Hungary and vice versa? Can we assume that the underlying problems are the same? Which knowledge will be circulated and define the policy agendas of the future? How will the limits between important knowledge and knowledge that remains on the margins be negotiated? These are questions likely to be considered however productive and rich the learning experiences within and across CoPs.

Community: CoPs in the neoliberal academy?

The second defining feature of a CoP concerns its "community" aspect. For a community to exist, there needs to be mutual engagement among its participants. "The community creates the social fabric of learning. A strong community fosters interactions and relationships based on mutual respect and trust. It encourages a willingness to share ideas, expose one's ignorance, ask difficult questions, and listen carefully" (Wenger et al., 2002, p. 28). Thus, what makes a community different from a group of employees who might belong to the same job category or a loose network of contacts is a habit of regular interaction which builds trusting relationships and a sense of belonging – not on any matter but on issues that are important to their domain. Despite the fact that a lot of effort and work is usually involved in cultivating a sense of community across diverse and contrasting views, tensions, or even conflict, there is tendency to conceive CoPs as a primarily harmonious, safe haven (Gherardi & Nicolini, 2000; Reynolds, 2000). Reaburn & McDonald (2017, p. 121), for example, suggest that CoPs provide precisely the means for "establishing collegial relations in a safe place that is free of hierarchical power and politics typically observed in schools and faculties". CoPs are frequently introduced as a space that lies orthogonal to the formal hierarchies and strategic priorities of organisations since the primary driver of a community is precisely a "shared interest" not governed by management but by self-interested, passionate individuals. Despite Lave & Wenger's (1991) initial recognition of the importance of power relations for learning communities, these issues have faded into the background in favour of a primary occupation for steering and managing self-organised communities. Wenger et al. (2002) speak in their later writings of "Cultivating Communities of Practice" (emphasis added), while Brown & Duguid (1991) popularise CoPs primarily as a "medium, and even as technology of consensus and stability" (Contu & Willmott, 2003, p. 284). Along these lines, many contributions in this book will confirm the pivotal role of the CoP facilitator for establishing and moving forward a CoP. Community in this sense involves a common history and shared identity, which does not imply that social relations are harmonious and tension-free.

Rather, the defining feature of the community lies in the "voluntary, informal and authentic" nature of its social relations which cannot be imposed because they are based upon authentic, personal interest and engagement (Wenger et al., 2002, p. 36).

The rather romantic account of community, however, is somehow at odds with the reality across contemporary HE institutions in Europe. Indeed, working conditions inside and outside the academia are less than favourable for establishing such safe spaces of togetherness. As Cox (2005, p. 533) writes, "... conditions of much, perhaps most twenty-first-century work inhibit sustained collective sense making, leading to fragmented, rather individualised appropriation of tasks". Specifically, feminist scholars have documented the pervasive and perverse effects of the "neoliberal university" (Slaughter & Rhoades, 2000) where contracts and careers have become more precarious while working demands have intensified. A new managerialism has cut funding and academic autonomy alike, requiring staff to do more with fewer resources and in less time: more teaching, more papers, more administrative committee work, more frequent reporting and engagement with (social) media (Anderson, 2008; Barry et al., 2001; Mountz et al., 2015; Ward, 2012). As Korczynski (2003) rightly observes, many CoPs in today's working environments resemble rather "communities of coping" than genuine opportunities for learning and emancipation.

Perversely, the speeding up of academic life towards output-oriented results goes hand in hand with the formation of a new regime of subjectivity that establishes new, subtle, internalised forms of self-control (Barker, 1993). Gill (2016, p. 42) observes how a new technology of the self is preoccupied with an endless task of "self-monitoring, planning, prioritising" which constitutes a "far more effective exercise of power than any imposed from above by employers". As a result, this perpetual process of self-optimisation is highly individualised and stands precisely in opposition to community building and collective action (Baker & Kelan, 2019; Pereira, 2016; Smidt et al., 2018; Vayreda et al., 2019). Often, the belief in meritocracy and individual choice in combination with increasing work demands effectively undermine the much-needed collective response, as it eliminates basically the possibility to recognise the structural foundation of precarious working conditions, including its built-in gender inequalities.

Translated into the context and experiences of the ACT project and its focus on gender equality, it is certainly true that CoPs provide an opportunity for community. Participants underline unanimously the advantages of overcoming one's isolation and connecting with others in similar, often marginalised positions within academic institutions. No doubt, resources are scarce in general and for gender equality, in particular, with inter-organisational CoPs offering the chance to pool assets and exchange experiences and strategies. The basis for some of the collaboration between institutions within the CoPs has in various instances been driven by informal networks of feminist activists/academics. However, it remains to be seen to what degree CoPs, despite their allure to "community" can activate a truly more collective and political mode of action. Examples of feminist collaboration and activism for a "slow" scholarship (Mountz et al., 2015; Pels, 2003), to the degree, that they do exist (Breeze & Taylor, 2020; O'Dwyer et al., 2018) manage without explicit references to CoPs.

At the policy level, however, it rather seems that CoPs might be misused and appropriated in the opposite direction, namely as a relatively "cheap" means to respond to the rising demand for gender equality work for example to access EU funding[4]. Unsurprisingly, it is women who carry the brunt of applying for example to the Athena SWAN certification in the United Kingdom (Caffrey et al., 2016; Ovseiko et al., 2017; Tzanakou & Pearce, 2019). Since gender equality work is primarily shouldered by women, there is a danger that associated responsibilities and tasks become an additional burden for those who should rather benefit from it. As Cox succinctly states along these lines, it is "at the very least, paradoxical to see how collaboration triggered by alienation can be turned into a management tool" (Cox, 2005, p. 533). From this critical angle, "community" becomes yet another means to embed employees more efficiently into organisational goals in order to fulfil "(reified) corporate objectives" (Rennstam & Kärreman, 2020). Even if the corporate objectives are laudable as in the case of gender equality, the responsibility and workload need to be distributed in a just manner.

Practice: Institutional change and alliances

The third and last defining feature of a CoP is "practice", already alluded to during the previous paragraphs. Albenga (2016) in her study of triggering structural change for gender equality in HE institutions highlights how, despite "awareness regarding the gendered biases of 'objective' excellence, these are left mostly unchallenged in practice". "Practice" is identified as the site where enacting "real" change happens. Along similar lines, Callerstig (2016, p. 119) reflects on transformational projects highlighting "the underlying assumption within this transformative idea is thus that a change in understanding can lead to a change in behaviour, and furthermore that change in individuals can lead to a change on an institutional level and impact existing policies and practices". These reflections highlight the primacy of focusing on "practice" in institutional change initiatives. The CoP literature has developed a sound body of knowledge reflecting on "practice" as knowledge and community through "doing" and "acting".

In their management-oriented book, Wenger and colleagues (2002, p. 38) emphasise that the primary task of a shared practice is to establish a "basic body of knowledge that creates a common foundation" which allows the members of the community to work together effectively. A "shared repertoire" or practice can include "routines, words, tools, ways of doing things, stories, gestures, symbols, genres, actions, or concepts" that crystallises past activities while providing the repertoire for its current and future activities (Wenger, 1998, p. 83). Although the practice is often understood as some sort of accomplished, output or solidified artefact of the community, it would be more accurate to conceive it as an activity. Practice implies "doings" such as shared behaviours and embodied understandings, including tacit conventions, subtle cues or well-tuned sensitivities (Wenger, 1998, p. 47).

The full implications of conceiving "practice" as "doing", as a truly process-oriented phenomenon that only exists to the extent that they are enacted, comes into view when consulting the critique of the CoP approach among organisation science scholars (Nicolini et al., 2003). In a rather radical comment on Wenger's work, Silvia Gherardi argues for the primacy of practice: instead of assuming "community" as the primary setting where learning takes place, we should rather consider how "situated and repeated actions create a context in which social relations among people, and between people and the material and cultural world, stabilise and become normatively sustained" (Gherardi, 2009, p. 523). Community is an effect of practice: it is through activities that a configuration of people, artefacts and social relations are held together and can form a joint enterprise and mutual engagement (Brown & Duguid, 1991; Nicolini, 2012; Roberts, 2006).

The notion that the world we inhabit is "routinely made and re-made in practice, using tools, discourse and our bodies" (Nicolini, 2017) immediately conjures up West & Zimmerman's classical essay on "Doing Gender" (1987) which conceives gender along the same lines, namely as a routine accomplishment embedded in everyday interaction. However, the "routine" aspect of action is only part of the story. Through continuous repetition, the social world including all its power relations, social injustice or gender inequality is made durable because it is inscribed in bodies and minds, tools and discourses and "knotted together in such a way that the results" of one inscription becomes the resources of another. The advantage of a close reading of the CoP literature along the lines of a practice-based approach to organisations should become clear: its emphasis on "doings" as well as their socio-material embedding facilitates the transition from a theory of social learning towards a much-needed understanding of organisational change (Bruni et al., 2004; Nicolini et al., 2003; Poggio, 2006). Changing social relations does not depend anymore on personal will, nor the generation of new insights and knowledge. Rather, it involves the rewiring of practice itself, which now means to decentre and transform the socio-material network that constitutes an academic organisation including its positions of privilege and marginality.

While the CoP literature often is largely limited by conceiving practice as an outcome of a CoP, a focus on gender equality and institutional change immediately conjures up a more complex picture. Even though CoPs will be more resilient and sustainable, the more they have established their own practice and identity, their overarching goal regarding gender equality needs to be seen in relation to more durable and solid practices of the embedding organisation. There is always a "nexus" (Nicolini, 2012) of competing practices where CoPs run in parallel, are co-opted, or are in open confrontation with wider established and emerging practices. One such emerging practice, for example, concerns the new politics of documentation introduced into HE which tends to reframe equality work as a bureaucratic exercise. Attached to existing procedures of quality control and accountability, the circulation of gender equality documents constitutes a practice which seems to supplant the actual equality work itself (Ahmed, 2007; Davis et al., 2010; Garforth & Kerr, 2009; Marx, 2019). From this perspective, focusing on "practice" then not only means building shared repertoire among CoP members but also understanding how one's own practice can possibly affect or re-enact these broader, gendered organisational requirements. It is through this development of alternative "doings" that the power of the CoP to de/en-gender organisational practices is unleashed and the role of the gender equality "practitioner" becomes paramount in the quest for institutional change.

The priority of practice has implications not only for our understanding of "community" but also for "knowledge/learning" and "power". Learning/knowledge generation means understanding how a concrete socio-material network is articulated. Gender equality practitioners, to the extent that they are always working in a specific time and place, within a specific organisation, have a deep understanding how organisational procedures, routines, forms of documentation and decision-making, unwritten rules or personal alliances interlock to produce "their" organisation. Practice in this sense implies a "site ontology", i.e. the primacy of a specific context for analysing and explaining social phenomena (Schatzki, 2005). The fact that generic insights, abstract theories, or even concrete examples from other times and places apply only to a limited degree is not surprising (Yanow, 2004): first, because each socio-material network is situated, constituting its site, but also because practice can never be reduced to words alone. Achieving structural change for greater gender equality requires power, now understood as shifting the "mundane practices of organizing" (Brown et al., 2010). Beyond words and intentions, it requires acting in such a way as to not re-enact established routines but enacting alternative practices, now conceived as the weaving of an alternative socio-material networks.

The full potential of a practice-based approach to CoPs, therefore, becomes visible when extending the concept to its wider organisational embedding. CoPs, to the degree that they are autonomous and thrive on the interest of their participants, provide the opportunity to explore alternative ways of "doings": a CoPs practice often exists in opposition or in parallel to existing organisational routines. They, therefore, become the experimental environment where alternatives can not only be (re)imagined and thought about but also put into practice – relatively unbound by existing organisational hierarchies and procedures. However, if CoPs are to become an instrument for advancing gender equality in contemporary academic organisations, then these CoP specific practices need to be extended and incorporated – mainstreamed – into the wider organisational environment. A CoP practice in this sense is never simply a means to produce knowledge – to the degree that knowledge only exists as practice, as something enacted and continuously re-enacted, it involves alliance building and drawing stakeholders and their resources in alternative processes of thinking and doing academic work. Knowledge creation is then inherently political and "slow" – as it involves forging alliances within organisations and across CoPs. Alliances that not only exist on paper but that have developed a shared practice, an alternative way of doing research, teaching and taking care of others. Members of CoPs should therefore always guard against being efficient or effective and insist on the autonomy to define their "shared concern" – which might or might not be aligned with the wider organisational agendas. To the degree that CoPs engage in gender equality work, their alternative practices can only accelerate change by slowly building and embedding alternative practices.

Overview of the chapters

This book provides a comprehensive overview of our experiences of setting up and supporting eight CoPs for gender equality in R&I and HE throughout Europe based on the ACT project. The empirical evidence has been gathered using various methodological approaches including participant observation, case studies and semi-structured interviews as well as a formal evaluation. During the writing process, two peer review sessions were held – where authors exchanged chapters and provided comments on another comparable chapter. This was followed by an open discussion – where all contributors to this book were able to comment and the authors then revised chapters. This approach proved fruitful in terms of fostering a common project resulting in increased synergies between the various chapters and thoughtful reflections throughout the book.

Individual chapters are distributed across three sections. In the first section, we aim to make advances on the conceptual and theoretical levels – crucially examining what a CoP approach can offer institutional change processes for a greater gender equality. This includes reflections on the main methods and tools designed to support CoPs such as the Gender Equality Audit and Monitoring (GEAM) tool as well as the co-creation methods toolkit. The second section containing Chapters 5–9 is predominantly developed by those contributors who were also CoP facilitators, often academics but also practitioners who have been responsible for the setting up and running of the CoPs. In the third section, comprising Chapters 10 and 11, the benefits and limits of a CoP approach to promoting gender equality in R&I is considered, followed by a reflection on its impact and effectiveness in terms of scaling up the approach regarding the three ERA objectives, namely careers, decision-making and integrating the gender dimension in teaching and research content.

In Chapter 2 of this volume, Thomson, Barnard, Hassan and Dainty provide a theoretical contribution which argues for developing a new concept – a Community of Political Practice (CoPP). They define a CoPP, as a group of institutionally affiliated people across different organisations or nations coalescing around a shared concern for social equality who engage in transformative practice, who learn from each other and co-create knowledge through regular interactions to act on institutional change. Despite the growth of CoP scholarship, theoretical explorations of CoPs designed for social or institutional change are scarce even though change can occur through peer collaboration and institutional work of embedded agents. This is a missed opportunity for institutional change efforts, and for furthering CoP theory. Thomson, Barnard, Hassan and Dainty argue that CoP is a promising mobilising structure for promoting equality endeavours drawing on the concepts of counter-hegemony, social movement and institutional change. By mapping out new crossroads of theories of CoP, social movement and institutional change, this chapter deepens insights into potential lessons to consider when designing CoPPs for counter-hegemonic endeavours.

Chapter 3 by Guyan, Aldercotte, Müller, Caprile and Yanes takes a more practical turn and examines the design process undertaken for the Gender Equality Audit and Monitoring (GEAM) tool. Developed by research teams from the United Kingdom (Advance HE) and Spain (Notus and Universitat Oberta de Catalunya), the GEAM provides a comprehensive, transferable and transnational survey for HE and research organisations that wish to undertake an audit of gender equality among academic, technical and support staff. The transnational roll-out of the survey has highlighted areas where ideas about gender equality, education and research and working practices are conceptualised differently. The chapter, therefore, provides an introduction for other research teams engaged in the design of equality, diversity and inclusion surveys; translation of surveys into multiple languages and used across multiple national contexts; navigation of challenges when they emerge; and use of a standardised framework to gather evidence of gender inequality across a range of thematic areas.

Chapter 4 by Thomson, Rabsch, Barnard, Hassan and Dainty is also a methods-based chapter and addresses a lack of specific ground-level tools and techniques for facilitators and community members involved in cultivating CoPs for institutional change. CoPs are complex and contextually sensitive social phenomena; thus, they require a facilitative framework to connect its members to co-create and collaborate. The chapter presents a selection of co-creation methods utilised in the ACT project and reflects how such methods enable CoPs to unleash their potential, as well as enables them to act as change agents towards institutional change. The chapter presents the backdrop of the CoP concept and its theoretical framework, the CoP definition, CoP lifecycle phases, as well as CoP success factors and primary areas of activity and argues how these theoretical elements provide a rationale for co-creation activities. Finally, the chapter considers in more detail four co-creation activities (DAKI Retrospective, 1-2-4-All, Plan of Change, and Future Workshop) to demonstrate their potential strengths in CoP collaboration efforts.

Chapter 5 by Sekula, Ciaputa, Warat, Krzaklewska, Beranek and Reidl opens the reporting on first-hand CoP experiences. Based upon the experience of the GEinCEE CoP, the authors examine to what extent CoPs can facilitate conditions for effective gender equality interventions in research and academia in Central and Eastern Europe. It also examines the usefulness of CoP as a mechanism to foster the necessary conditions for advancing gender equality. These include: the agency of change actors; the engagement of organisational stakeholders; building up gender know-how; access to practical tools for designing evidence-based interventions; managing resistance; framing gender equality within wider concepts and human and financial resources.

Chapter 6 developed by Mihajlović Trbovc then goes on to describe the CoP for Alternative Infrastructure for Gender Equality in Academic Institutions (Alt+G) that brings together researchers from Slovenia. Building on a history of efforts to achieve gender equality in Slovene academia, the CoP gathers researchers (and some academic staff) dedicated to promoting women in science, improving gender equality in their institutions and the sector as a whole. The chapter demonstrates how the focus of transformative efforts shifted from the level of national regulations to the academic institutions, due to systemic conditions. Furthermore, it shows that the CoP approach is particularly beneficial for spreading and multiplying structural change within HE institutions and research organisations, and that it can help overcome certain systemic fallacies. The CoP structure and sense of community is able to provide a framework that turns unforeseen challenges into windows of opportunity for institutional change and creates space for mutual learning. Since the CoP approach operates on the fuel of personal motivation and depends on individual rather than institutional commitment, its ability and reach in enhancing concrete institutional change is contingent on favourable structural context.

Chapter 7 reflects on how disciplinary-specific CoPs can be a useful vehicle to share knowledge, experience and practices to further gender equality in R&I organisations. Reiland and Kamlade share their experiences of setting up two CoPs with specific disciplinary focuses, one on physics (GENERA) and the other on life sciences (LifeSciCoP). There is a dearth of academic literature that looks at how disciplinary based CoPs can foster institutional change for gender equality in R&I institutions. By charting the similarities and differences of their approaches to sharing knowledge, experience and practices for gender equality, important insights emerge on how disciplinary context factors shape the CoP approach and provide entry points for gender equality work.

Chapter 8 by Axelsdóttir, Steinþórsdóttir and Einarsdóttir reflects on the opportunities and obstacles of CoPs in developing and implementing gender budgeting to challenge gender biases in decision-making of research performing organisations. Through "Targeted Implementation Projects", the CoP aims to develop shared knowledge on how to implement gender budgeting in order to further the objective of gender equality in decision-making within RPOs. Drawing on a case study based on the GenBUDGET CoP, which includes 21 representatives in 14 RPOs, the analysis explores the potential for an international CoP to harness inter-organisational cooperation and create knowledge about gender budgeting when CoP members' knowledge about that strategy is very diverse from the outset.

Chapter 9 by Damala, Mour and Godfroy presents the underlying motivations and inner workings of the Strategies for Sustainable Gender Equality CoP. They provide an overview of how and why STRATEGIES was launched, offering both an empirical as well as an experiential account of what has been achieved, bringing into the picture conceptual, theoretical and practical underpinnings from the life of the CoP, from its inception to the end of the ACT project and its transitioning to a new network. The focal point and interest of the collaboration initiated by STRATEGIES is "sustainability", a concept defined in relation to recent developments in sustainable development as well as in project management. They describe the philosophy, methodology and all concrete steps they have followed to set up and launch the CoP, and stimulate the exchange of knowledge, policies, know-how and lessons learned both onsite – as well as in the post-COVID-19 world – online.

Chapter 10 opens the third and final section of the book. Reidl, Baranek and Holzinger investigate the added value of CoPs for the implementation of gender equality strategies for their members and member organisations based on the evaluation carried out during the ACT project. By applying Wenger's concept of value creation (Wenger et al, 2011) they demonstrate which different values and benefits are created through participating in CoPs. Based on interview data and self-reporting data across seven CoPs, they show the immediate, potential and applied values of CoP participation. Members were seen to benefit from CoPs in many ways, for example, through new contacts, new knowledge, empowerment, active implementation support and much more. In addition to an analysis of the added value of CoPs, in this chapter, the authors explore the question of whether the added value that has been identified is sufficient to promote gender equality in research performing and research funding organisations, or whether additional activities are needed to achieve this. They also reflect on the limitations of the CoP approach to institutional change towards gender equality.

Chapter 11 highlights how knowledge sharing between and beyond the CoPs has been achieved specifically in relation to the three ERA objectives for gender equality and mainstreaming: careers, decision-making and integrating the gender dimension in teaching and research content. In scaling up the CoP approach, ACT established three so-called ERA priority coordination groups which identified and addressed cross-cutting issues related to each of the objectives. These groups brought together ACT Consortium partners, members of different CoPs, ACT advisory board members, experts, representatives of ERA level players and other relevant R&I representatives from the CoPs' contexts – including local, regional, national and disciplinary networks. Through collaborative working, and the sharing of cutting-edge good practices, each of these groups have made substantive contributions to the debate on how to make progress in each of these areas. This chapter details the main debates in each of these three areas and tries to shed light on the priorities for future collaborative work.

The concluding chapter by Rachel Palmén and Jörg Müller revisits the conceptual issues outlined in the introduction in the light of the individual chapters and spells out some of the implications in view of the wider CoP and gender equality literature.

Notes

1. "Communities of PrACTice for Accelerating Gender Equality and Institutional Change in Research and Innovation across Europe" Horizon 2020 project, grant number 788204 is referred to throughout this book as "The ACT project". See also https://www.act-on-gender.eu.
2. The EU-15 countries include: Austria, Belgium, Denmark, Finland, France, Germany, Greece, Ireland, Italy, Luxembourg, Netherlands, Portugal, Spain, Sweden and United Kingdom.
3. The EU-13 countries include: Bulgaria, Croatia, Cyprus, Czech Republic, Estonia, Hungary, Latvia, Lithuania, Malta, Poland, Romania, Slovakia and Slovenia.
4. See GEPs as eligibility criterion for accessing Horizon Europe funding (European Commission, 2021a).

References

Ahmed, Sara (2007). 'You End Up Doing the Document Rather Than Doing the Doing': Diversity, Race Equality and the Politics of Documentation. *Ethnic and Racial Studies*, *30*(4), 590–609. https://doi.org/10.1080/01419870701356015

Albenga, Viviane (2016). Between Knowledge and Power: Triggering Structural Change for Gender Equality from Inside in Higher Education Institutions. In M. Bustelo, L. Ferguson, & M. Forest (Eds.), *The Politics of Feminist Knowledge Transfer: Gender Training and Gender Expertise* (pp. 139–156). Palgrave Macmillan. https://doi.org/10.1057/978-1-137-48685-1

Amin, Ash, & Roberts, Joanne (2006). Communities of practice? Varieties of situated learning. A paper prepared for: EU Network of Excellence Dynamics of Institutions and Markets in Europe (DIME).

Anderson, Elizabeth (1995). Feminist Epistemology: An Interpretation and a Defense. *Hypatia*, *10*(3), 50–84. https://doi.org/10.1111/j.1527-2001.1995.tb00737.x

Anderson, Gina (2008). Mapping Academic Resistance in the Managerial University. *Organization*, *15*(2), 251–270. https://doi.org/10.1177/1350508407086583

Baker, Darren T., & Kelan, Elisabeth K. (2019). Splitting and Blaming: The Psychic Life of Neoliberal Executive Women. *Human Relations, 72*(1), 69–97. https://doi.org/10.1177/0018726718772010

Barker, James A. (1993). Tightening the Iron Cage – Concertive Control in Self-Managing Teams. *Administrative Science Quarterly, 38*(3), 408–437.

Barnard, Sarah, Hassan, Tarek, Dainty, Andrew, Polo, Lucia, Arrizabalaga, Ezekiela (2016*). Using communities of practice to support the implementation of gender equality plans: lessons from a cross-national action research project*. Loughborough University. Conference contribution. https://hdl.handle.net/2134/23681

Barry, Jim, Chandler, John, & Clark, Heather (2001). Between the Ivory Tower and the Academic Assembly Line*. *Journal of Management Studies, 38*(1), 88–101. https://doi.org/10.1111/1467-6486.00229

Blackmore, Chris (Ed.). (2010). *Social Learning Systems and Communities of Practice.* Springer.

Bolisani, Ettore, & Scarso, Enrico (2014). The Place of Communities of Practice in Knowledge Management Studies: A Critical Review. *Journal of Knowledge Management, 18*(2), 366–381. https://doi.org/10.1108/JKM-07-2013-0277

Breeze, Maddie, & Taylor, Yvette (2020). Feminist Collaborations in Higher Education: Stretched across Career Stages. *Gender and Education, 32*(3), 412–428. https://doi.org/10.1080/09540253.2018.1471197

Brown, Joh S., & Duguid, Paul (1991). Organizational Learning and Communities-of-Practice: Toward a Unified View of Working, Learning, and Innovation. *Organization Science, 2*(1), 40–57. https://doi.org/10.1287/orsc.2.1.40

Brown, Andrew D., Kornberger, Martin, Clegg, Stewart R., & Carter, Chris (2010). 'Invisible Walls' and 'Silent Hierarchies': A Case Study of Power Relations in an Architecture Firm. *Human Relations, 63*(4), 525–549. https://doi.org/10.1177/0018726709339862

Bruni, Attile, Gherardi, Silvia, & Poggio, Barbara (2004). Doing Gender, Doing Entrepreneurship: An Ethnographic Account of Intertwined Practices. *Gender, Work and Organization, 11*(4), 406–429. https://doi.org/10.1111/j.1468-0432.2004.00240.x

Caffrey, Louise, Wyatt, David, Fudge, Nina, Mattingley, Helena, Williamson, Catherine, & McKevitt, Christopher (2016). Gender Equity Programmes in Academic Medicine: A Realist Evaluation Approach to Athena SWAN Processes. *BMJ Open, 6*(9), e012090. https://doi.org/10.1136/bmjopen-2016-012090

Callerstig, Anne-Charlott (2016). Gender Training as a Tool for Transformative Gender Mainstreaming: Evidence from Sweden. In M. Bustelo, L. Ferguson, & M. Forest (Eds.), *The Politics of Feminist Knowledge Transfer: Gender Training and Gender Expertise* (pp. 118–138). Palgrave Macmillan. https://doi.org/10.1057/978-1-137-48685-1

Contu, Alessia, & Willmott, Hugh (2003). Re-Embedding Situatedness: The Importance of Power Relations in Learning Theory. *Organization Science, 14*(3), 283–296. https://doi.org/10.1287/orsc.14.3.283.15167

Council of the European Union. (2015). *Advancing Gender Equality in the European Research Area– Council Conclusions (adopted on 01/12/2015)* (No. 14846/15). Council of the European Union. https://data.consilium.europa.eu/doc/document/ST-14846-2015-INIT/en/pdf

Cox, Andrew (2005). What Are Communities of Practice? A Comparative Review of Four Seminal Works. *Journal of Information Science, 31*(6), 527–540. https://doi.org/10.1177/0165551505057016

Curnow, Joe (2013). Fight the Power: Situated Learning and Conscientisation in a Gendered Community of Practice. *Gender and Education, 25*(7), 834–850. https://doi.org/10.1080/09540253.2013.845649

Davis, Kevin E., Kingsbury, Benedict, & Merry, Sally E. (2010). Indicators as a Technology of Global Governance. *NYU Law and Economics Research Paper, 10–13.* http://papers.ssrn.com/abstract=1583431

Dubé, Line, Bourhis, Anne, & Jacob, Réal (2003). Towards a Typology of Virtual Communities of Practice. *Interdisciplinary Journal of Information, Knowledge, and Management, 1,* 3–13.

ERAC Standing Working Group on Gender in Research and Innovation. (2018). *Report on the implementation of Council Conclusions of 1 December 2015 on Advancing Gender Equality in the European Research Area* (No. 1213/18). European Research Area and Innovation Committee. https://data.consilium.europa.eu/doc/document/ST-1213-2018-INIT/en/pdf

European Commission. (2011). *Structural Change in Research Institutions: Enhancing Excellence, Gender Equality and Efficiency in Research and Innovation.* Publications Office of the European Union. https://doi.org/10.2777/94044

European Commission. (2020). *A New ERA for Research and Innovation: Commission Staff Working Document* (Commission Staff Working Document COM (2020) 628). Publications Office of the European Union. https://data.europa.eu/doi/10.2777/605834

European Commission. (2021a). *Horizon Europe Guidance on Gender Equality Plans.* Publications Office of the European Union. https://data.europa.eu/doi/10.2777/876509

European Commission. (2021b). *She figures 2021: Gender in Research and Innovation: Statistics and Indicators.* Publications Office of the European Union. https://data.europa.eu/doi/10.2777/06090

European Commission, Tornasi, Z., & Delaney, N. (2020). *Gender Equality: Achievements in Horizon 2020 and Recommendations on the Way Forward.* Publications Office of the European Union. https://data.europa.eu/doi/10.2777/009204

Ferguson, Lucy (2021). *Analytical Review. Structural Change for Gender Equality in Research and Innovation* (No. 4). Ministry of Education and Culture. http://urn.fi/URN:ISBN:978-952-263-880-9

Garforth, Lisa, & Kerr, Anne (2009). Women and Science. What's the Problem? *Social Politics,* 379–403.

Gherardi, Silvia (2009). Community of Practice or Practices Of a Community. In S. J. Armstrong & C. V. Fukami (Eds.), *The Sage Handbook of Management Learning, Education, and Development* (pp. 514–530). SAGE Publications Ltd. https://books.google.es/books?id=Om3nZSDGKNUC

Gherardi, Silvia, & Nicolini, Davide (2000). To Transfer is to Transform: The Circulation of Safety Knowledge. *Organization, 7*(2), 329–348. https://doi.org/10.1177/135050840072008

Gill, Rosalind (2016). Breaking the Silence: The Hidden Injuries of Neo-Liberal Academia. *Feministische Studien, 34*(1), 39–55. https://doi.org/10.1515/fs-2016-0105

Hara, Noriko, Shachaf, Pnina, & Stoerger, Sharon (2009). Online Communities of Practice Typology Revisited. *Journal of Information Science, 35*(6), 740–757. https://doi.org/10.1177/0165551509342361

Haraway, Donna (1988). Situated Knowledges: The Science Question in Feminism and the Privilege of Partial Perspective. *Feminist Studies, 14*(3), 575–599. https://doi.org/10.2307/3178066

Harding, Sandra (1986). *The Science Question in Feminism.* Cornell University Press.

Harding, Sandra (2004). *The Feminist Standpoint Theory Reader: Intellectual and Political Controversies.* Routledge.

Holmes, Janet, & Meyerhoff, Miriam (1999). The Community of Practice: Theories and Methodologies in Language and Gender Research. *Language in Society, 28*(2), 173–183. https://doi.org/10.1017/S004740459900202X

Intemann, Kristen (2010). 25 Years of Feminist Empiricism and Standpoint Theory: Where Are We Now? *Hypatia, 25*(4), 778–796. https://doi.org/10.1111/j.1527-2001.2010.01138.x

Janssens, Maddy, & Zanoni, Patrizia (2021). Making Diversity Research Matter for Social Change: New Conversations Beyond the Firm. *Organization Theory, 2*(2), 26317877211004604. https://doi.org/10.1177/26317877211004603

Korczynski, Marek (2003). Communities of Coping: Collective Emotional Labour in Service Work. *Organization, 10*(1), 55–79. https://doi.org/10.1177/1350508403010001479

Lave, Jean, & Wenger, Etienne (1991). *Situated Learning: Legitimate Peripheral Participation.* Cambridge University Press.

Longino, Helene E. (1990). Science as Social Knowledge. In *Science as Social Knowledge.* Princeton University Press. https://www.degruyter.com/document/doi/10.1515/9780691209753/html

Marx, Ulrike (2019). Accounting for equality: Gender budgeting and moderate feminism. *Gender, Work & Organization, 26*(8), 1176–1190. https://doi.org/10.1111/gwao.12307

McDonald, Jacquie, & Cater-Steel, Aileen (Eds.). (2017). *Communities of Practice. Facilitating Social Learning in Higher Education.* Springer. https://link.springer.com/book/10.1007/978-981-10-2879-3

Mountz, Alison, Bonds, Anne, Mansfield, Becky, Loyd, Jenna, Hyndman, Jennifer, Walton-Roberts, Margaret, Basu, Ranu, Whitson, Risa, Hawkins, Roberta, Hamilton, Trina, & Curran, Winifred (2015). For Slow Scholarship: A Feminist Politics of Resistance through Collective Action in the Neoliberal University. *ACME: An International Journal for Critical Geographies, 14*(4), 1235–1259.

Nicolini, Davide, Gherardi, Silvia, & Yanow, Dvora (Eds.). (2003). *Knowing in Organizations: A Practice-Based Approach.* M.E. Sharpe.

Nicolini, Davide (2012). *Practice Theory, Work, and Organization: An Introduction.* OUP Oxford.

Nicolini, Davide (2017). Practice Theory as a Package of Theory, Method and Vocabulary: Affordances and Limitations. In M. Jonas, B. Littig, & A. Wroblewski (Eds.), *Methodological Reflections on Practice Oriented Theories* (pp. 19–34). Springer International Publishing. https://doi.org/10.1007/978-3-319-52897-7_2

O'Dwyer, Siobhan, Pinto, Sarah, & McDonough, Sharon (2018). Self-Care for Academics: A Poetic Invitation to Reflect and Resist. *Reflective Practice, 19*(2), 243–249. https://doi.org/10.1080/14623943.2018.1437407

Ovseiko, Pavel V., Chapple, Alison, Edmunds, Laurel D., & Ziebland, Sue (2017). Advancing Gender Equality through the Athena SWAN Charter for Women in Science: An Exploratory Study of Women's and Men's Perceptions. *Health Research Policy and Systems, 15*(1), 12. https://doi.org/10.1186/s12961-017-0177-9

Paechter, Carrie (2003). Masculinities and Femininities as Communities of Practice. *Women's Studies International Forum, 26*(1), 69–77. https://doi.org/10.1016/S0277-5395(02)00356-4

Pattinson, Steven, Preece, David, & Dawson, Patrick (2016). In Search of Innovative Capabilities of Communities of Practice: A Systematic Review and Typology for Future Research. *Management Learning, 47*(5), 506–524. https://doi.org/10.1177/1350507616646698

Pels, Dick (2003). Unhastening Science: Temporal Demarcations in the 'Social Triangle'. *European Journal of Social Theory, 6*(2), 209–231. https://doi.org/10.1177/1368431003006002004

Pereira, Maria do Mar (2016). Struggling within and beyond the Performative University: Articulating Activism and Work in an "Academia without Walls". *Women's Studies International Forum, 54*, 100–110. https://doi.org/10.1016/j.wsif.2015.06.008

Poggio, Barbara (2006). Editorial: Outline of a Theory of Gender Practices. *Gender, Work and Organization, 13*(3), 225–233. https://doi.org/10.1111/j.1468-0432.2006.00305.x

Pyrko, Igor, Dörfler, Victor, & Eden, Colin (2019). Communities of Practice in Landscapes of Practice. *Management Learning, 50*(4), 482–499. https://doi.org/10.1177/1350507619860854

Reaburn, Peter, & McDonald, Jacquie (2017). Creating and Facilitating Communities of Practice in Higher Education: Theory to Practice in a Regional Australian University. In J. McDonald & A. Cater-Steel (Eds.), *Communities of Practice. Facilitating Social Learning in Higher Education* (pp. 121–150). Springer. https://link.springer.com/book/10.1007/978-981-10-2879-3

Rennstam, Jens, & Kärreman, Dan (2020). Understanding Control in Communities of Practice: Constructive Disobedience in a High-Tech Firm. *Human Relations, 73*(6), 864–890. https://doi.org/10.1177/0018726719843588

Reynolds, Michael (2000). Bright Lights and the Pastoral Idyll: Ideas of Community Underlying Management Education Methodologies. *Management Learning, 31*(1), 67–81. https://doi.org/10.1177/1350507600311006

Roberts, Joanne (2006). Limits to Communities of Practice. *Journal of Management Studies, 43*(3), 623–639. https://doi.org/10.1111/j.1467-6486.2006.00618.x

Schatzki, Theodore R. (2005). Peripheral Vision: The Sites of Organizations. *Organization Studies, 26*(3), 465–484. https://doi.org/10.1177/0170840605050876

Schulte, Benjamin (2020). *The Organizational Embeddedness of Communities of Practice: Exploring the Cultural and Leadership Dynamics of Self-organized Practice.* Springer Nature. https://books.google.es/books?id=AwgLEAAAQBAJ

Slaughter, Sheila, & Rhoades, Gary (2000). The Neo-Liberal University. *New Labor Forum, 6*, 73–79.

Smidt, Thomas B., Bondestam, Fredrik, Pétursdóttir, Gyða Margrét, & Einarsdóttir, Þorgerður (2018). Expanding Gendered Sites of Resistance in the Neoliberal Academy. *European Journal of Higher Education, 10*(2), 115–129. https://doi.org/10.1080/21568235.2018.1541753

Stapleton, Karyn (2001). Constructing a Feminist Identity: Discourse and the Community of Practice. *Feminism & Psychology*, *11*(4), 459–491. https://doi.org/10.1177/0959353501011004003

Terry, Daniel R., Nguyen, Hoang, Peck, Blake, Smith, Andrew, & Phan, Hoang (2020). Communities of Practice: A Systematic Review and Meta-Synthesis of What It Means and How It Really Works among Nursing Students and Novices. *Journal of Clinical Nursing*, *29*(3–4), 370–380. https://doi.org/10.1111/jocn.15100

Tzanakou, Charikleia, & Pearce, Ruth (2019). Moderate Feminism within or against the Neoliberal University? The Example of Athena SWAN. *Gender, Work & Organization*, *26*(8), 1191–1211. https://doi.org/10.1111/gwao.12336

Vayreda, Agnès, Conesa, Ester, Revelles-Benavente, Beatriz, & Ramos, Ana M. G. (2019). Subjectivation Processes and Gender in a Neoliberal Model of Science in Three Spanish Research Centres. *Gender, Work & Organization*, *26*(4), 430–447. https://doi.org/10.1111/gwao.12360

Wagner, Ina (1994). Connecting Communities of Practice: Feminism, Science, and Technology. *Women's Studies International Forum*, *17*(2), 257–265. https://doi.org/10.1016/0277-5395(94)90032-9

Ward, Steven C. (2012). Neoliberalism and the Global Restructuring of Knowledge and Education. In *Routledge, Taylor & Francis Group*. Routledge, Taylor & Francis Group. https://books.google.es/books?hl=es&lr=&id=6ujGBQAAQBAJ

Wenger, Etienne, Trayner, Beverly, de Laat, Maarten (2011). Promoting and assessing value creation in communities and networks: A conceptual framework. Ruud de Moor Centrum https://www.researchgate.net/publication/220040553_Promoting_and_Assessing_Value_Creation_in_Communities_and_Networks_A_Conceptual_Framework

Wenger-Trayner, Etienne, Fenton-O'Creevy, Mark, Hutchinson, Steven, Kubiak, Chris, & Wenger-Trayner, Beverly (Eds.). (2014). *Learning in Landscapes of Practice: Boundaries, identity, and knowledgeability in practice-based learning.* Routledge.

Wenger, Etienne (1998). *Communities of Practice: Learning, Meaning, and Identity.* Cambridge University Press. https://books.google.es/books?id=heBZpgYUKdAC

Wenger, Etienne, McDermott, Richard A., & Snyder, William (2002). *Cultivating Communities of Practice: A Guide to Managing Knowledge.* Harvard Business Press. https://books.google.es/books?id=m1xZuNq9RygC

West, Candace, & Zimmerman, Don H. (1987). Doing Gender. *Gender & Society*, *1*(2), 125–151. https://doi.org/10.1177/0891243287001002002

Wylie, Alison (2003). Why Standpoint Matters. In R. Figueroa & S. G. Harding (Eds.), *Science and other cultures: Issues in philosophies of science and technology* (pp. 26–48). Routledge.

Yanow, Dvora (2004). Translating Local Knowledge at Organizational Peripheries*. *British Journal of Management*, *15*(S1), 9–25. https://doi.org/10.1111/j.1467-8551.2004.t01-1-00403.x

2 Can a Community of Practice foster institutional change for gender equality?

Conceptualising 'Community of Political Practice'

Aleksandra Thomson, Sarah H. Barnard, Tarek M. Hassan and Andrew Dainty

Introduction

This chapter positions Community of Practice (CoP) as an approach to organising for institutional change for gender equality in research and innovation (R&I), drawing on theories of counter-hegemony, social movement, institutional change, and CoP to propose a new concept of Community of Political Practice (CoPP). The intention is to move beyond the focus on situated and peer learning within CoPs, and illuminate how CoPPs – underpinned by transformative objectives – could contribute to gender equality efforts in institutions with skewed distributions of power and resources. We do not proclaim that this is a demonstrably effective way of delivering change; rather, we offer a conceptual starting point for a new understanding of CoPs.

CoPs can be either organically or purposefully organised to collectively learn and generate knowledge about a shared domain (Iaquinto, Ison, & Faggian, 2011; Wenger, McDermott, & Snyder, 2002). CoP has been broadly defined as a group of people who coalesce around a particular concern or passion and learn how to improve their practices and know-how as they regularly interact (Wenger et al., 2002). Learning occurs through doing (practice), belonging (community), becoming (identity), and meaning (experience) (Wenger, 1998). Much of CoP-focused research concentrates on the processes of knowing and learning in action underpinned by managerial applications (Amin & Roberts, 2008); innovation (Swan, Scarbrough, & Robertson, 2002); higher education (HE) curriculum (Annala & Mäkinen, 2017; Goodyear, Casey, & Kirk, 2016); or teaching as a social practice (Kendall et al., 2018).

While work in this vein has offered important insights, organising through CoPs to facilitate institutional change has been neglected by both social movement and CoP scholarship, with exceptions pertaining to how CoPs face

DOI: 10.4324/9781003225546-2

institutional change (Hutchins & Boyle, 2017), or how change of practice is highly political (Mørk et al., 2010). We argue that this is a missed opportunity for change efforts (e.g., promoting gender equality in R&I), and for furthering CoP theory. More importantly, examining the crossroads of CoP and social movement theory could underscore potential lessons from the two bodies of knowledge if CoPs with institutional change-focused domains are to be effective. Institutional change through innovative or alternative practice and embedded actions is a political endeavour (Mørk et al., 2010; Drazin, 1990). Thus, this chapter aims to reconceptualise CoP as an approach for promoting institutional change through the development of the concept of *CoPP*, here defined as *a group of institutionally affiliated people across different organisations or nations coalescing around a shared concern for social equality who engage in transformative practice, who learn from each other and co-create knowledge through regular interactions to act on institutional change.* By political we mean relating to 'a quest for resource control and institutional legitimacy' (Mørk et al., 2010, p. 588) through the level of gender-equality concerned practice and embedded actions, with *the matter of concern* (Latour, 2007) being gender inequality.

We do so partly inspired by the concept of counter-hegemony (Gramsci, 1985; 1971) that contests and tries to disrupt normative views about social and political reality. This process may be achieved through a contemporary social movement as 'its actions challenge or break the limits of a system of social relations' (Melucci, 1989, p. 38). By consolidating resources and acting outside established structures, interest groups and movements can create independent organisational bases for advancing alternatives. Furthermore, CoPs with domains focused on institutional change and designed to be both inter- and intra- organisational could accelerate the process of change as they have the potential to start contestations from within (Carroll & Ratner, 1994). In addition, the conditions for institutional change are enhanced through links between social movements and communities that define their identity through a common goal (Gläser, 2004). Participants act within a social movement because they feel they share their intention to bring about change together with other members. To bring together these ideas and conceptualise Communities of PrACTice for Accelerating Gender Equality and Institutional Change in Research and Innovation across Europe (ACT), which form the focus of this volume, we also draw on key elements of organisations (Ahrne and Brunsson, 2010). This is because organisations provide bases for alternatives and drawing on these considerations, we complement the well-explored situated learning and knowing-in-action view of CoP, with the underexplored area of CoP efforts towards institutional change for gender equality in R&I.

Institutional change for gender equality in R&I

In the ACT project, institutional change for gender equality in R&I is defined as the promotion of an institutional environment (such as values, norms, structures and procedures) where gender equality is widely discussed and explicitly embraced in institutional and individual practices, decision making, and education, research and innovation content. To understand institutional change in the context of gender dominance and power dynamics, we draw on historical institutionalism to view institutions as legacies of historical struggles (Waylen, 2014). Institutions are collections of procedures and structures that both define and defend particular interests, and thus are 'political actors in their own right' (Conran & Thelen, 2016, p. 53). Institutional rules, norms, and practices shape power relations with unequal distributional consequences, awarding resources and concessions to those with power and disadvantaging others. Yet, the repercussions of differential power distribution galvanise desire for change (Waylen, 2014). Institutions are contested because they instantiate power. However, those who have initially been hindered endure, circumvent and subvert rules, and also start to occupy and redeploy institutions that have been designed and dominated by others (Conran & Thelen, 2016).

Gender is one of many cultural categories of difference implicated in the understructure of organisations (Acker, 1998) regulating who occupies positions of power. Despite societal changes and the reclaiming of women's history, the most important institutions such as the law, politics, religion, the state, and the economy are still dominated by white men (Acker, 1998). The institution of academy is also dominated by white masculine elites, 'historically developed by men, currently dominated by men, and symbolically interpreted from the standpoint of men in leading positions, both in the present and historically' (Acker, 1992, p. 567). Thus far, white men have benefitted from being in positions of power in greater numbers than women or people of other characteristics in HE and R&I, and as such, gender inequality persists in all aspects of R&I in Europe (Ferguson, 2021; Higher Education Statistics Agency, 2020; European Commission, 2019; Eurostat, 2019). Furthermore, policies that focused on trying to improve the situation by 'adding women and stirring' failed to produce a significant shift towards gender equality in these institutions due to gender bias and deeply ingrained cultural norms carried forward from the past (Ferguson, 2021; Mahoney and Thelen, 2010; Cotterill, Jackson, & Letherby, 2007).

There is also limited evidence that the 'critical mass' idea is effective (Grey, 2006). This is based on the premise that when the number of women in a given context reaches a certain level, e.g., 30%, issues of isolation, tokenism, and paucity of role models are reduced or removed. For instance, there have been suggestions that 'critical acts' (Lovenduski, 2001; Dahlerup, 1988) or 'safe spaces' (Childs, 2004) are more important for the substantive representation of women. Grey (2006) suggests that different sizes/percentages of critical masses may be needed, depending on the sought outcome.

Consequently, in 2011, the European Commission initiated a shift in focus away from 'fixing the women' towards 'addressing the structural transformation of institutions, using a systemic, comprehensive and sustainable approach', as specific initiatives designed to help women scientists had proved insufficient for addressing the 'leaky pipeline' (cited in Ferguson, 2021, p. 13). The focus on institutional change raises significant questions for gender equality practitioners, as there are many forms of challenges and changes to rules, norms, and procedures in institutions, and they are dependent on the extent to which these rules can be contested, as well as re-interpreted and re-implemented (Waylen, 2014). Rule contestation opportunities can arise through changing institutional contexts, such as changes to social and cultural expectations. For example, HE institutions facing pressures for change – from government, society, marketisation, the caprices of institutional leadership, and institutional restructuring, which can open space for re-interpretations (Conran & Thelen, 2016). Nevertheless, change is often gradual and endogenous if the status quo is defended by those in power as they have higher veto and discretion possibilities than the challengers (Waylen, 2014).

In the context of gender equality in R&I and HE, it is expected that the status quo will be defended by its beneficiaries and those who are assigned disproportionately more power, i.e., white male elites (Teelken & Deem, 2013). Indeed, opaqueness in decision-making and implicit bias found in assessment of merit, excellence, suitability for leadership, and performance evaluation have been identified as the main barriers to achieving gender equality in R&I at the institutional level (European Commission, 2012). Moreover, gender expertise within those institutions is also chronically underestimated, denied, and unacknowledged (Albenga, 2016). However, the changing institutional context which increasingly recognises an urgent need for gender equality actions (such as European Commission funding structures, Athena SWAN) fertilises the ground for gradual change and softening of institutional rules, their interpretation and enforcement. This change must be strategically planned to be effective (Waylen, 2014) and change agents have to acknowledge the importance of process, participation, and reflexivity throughout the life cycle of change interventions, rather than solely focusing on implementation outcomes (Ferguson, 2021).

Institutional change could also occur through 'institutional isomorphism' (DiMaggio & Powell, 1995), a norm once adopted in an institutional setting, spreading rapidly to a wide range of other institutions whose members accept the legitimacy of that norm (Hafner-Burton & Pollack, 2002, p. 340). This phenomenon is partly related to the swift and nearly universal adoption of gender mainstreaming by European governmental, and key international organisations (Hafner-Burton & Pollack, 2002), albeit not without the critique that technocratic approaches can depoliticise change (Verge, Ferrer-Fons, & González, 2018; Lombardo & Meier, 2006). For example, institutional applications for the gender charter award Athena SWAN might stem from cultural expectations, attempting to enhance legitimacy and attractiveness to the potential pool of students and staff (Barnard, 2017), yet this charter has been criticised as a tick-box exercise and a form of 'moderate feminism' (Tzanakou and Pearce, 2019). However, institutions might also be 'forced' into (coercive) isomorphism, as for instance, between 2011 and 2020, the National Institute for Health Research (NIHR) in the United Kingdom required a silver Athena SWAN award as a condition of funding. This has since been dropped after the government's call for reduction in bureaucracy, and the decision decried for particularly bad timing (COVID-19 impact on female academic careers) (McIntyre, 2020). This indicates that even seemingly radical, sudden, sweeping institutional change efforts can be precarious and volatile.

Mahoney and Thelen (2010) propose four types of institutional change: displacement (removing old rules), layering (creating new rules alongside old, e.g., quotas), drift (the impact of old rules loses significance), and conversion (leveraging rule ambiguity from within the system). Waylen (2014) argues that layering and conversion are best suited for gender equality. This is because these strategies are gradual and internally driven and are thus the most widely used and they can arguably be more immune to volatility of external forces. Layering and conversion may be more successful if change agents do not possess enough power to displace existing rules, yet they have enough power to create new rules or use existing rules in innovative ways. This is relevant for transnational equality projects designed for institutional change in several ways.

Firstly, layering requires insider knowledge of the existing rules, norms, and practices within the institution to develop a thorough understanding of how new rules can be slotted into the existing system. Powerful status quo players can protect old institutions, but they cannot prevent new elements from being added, which can hopefully tilt the balance towards the latter in the long run (Mahoney and Thelen, 2010). CoPs could be set up to coalesce a variety of members across different institutions and organisations transnationally to appreciate the diversity of approaches. Members deeply situated in local contexts would be able to leverage internal knowledge about 'how to go on' in their institutions to attempt to change structures (Stones, 2005). This is particularly relevant to the gender equality plan (GEP) requirement for EC grants, which becomes the 'new rule' layered alongside other funding avenues, and thus institutions need to understand, accept and possibly incorporate the new rules into their research grant application strategy.

Secondly, conversion also requires knowledge about existing rules within an institution, so they can be strategically redeployed in new ways. Change agents need to exploit the inherent ambiguities of institutions and re-interpret the rules to fit the change agenda (Waylen, 2014). Through conversion, the rules, norms, and practices are redirected to new goals, functions, or purposes, for instance by deploying existing resources to new ends (Streeck & Thelen, 2005). CoP as a structure for innovation opportunities is particularly suited to this approach (Bridwell-Mitchell, 2016; Brown & Duguid, 1991). An example of conversion is gender budgeting, which utilises existing resource allocation procedures in a gender equality framework (O'Hagan & Klatzer, 2018).

We will now shift our discussion to counter-hegemony and social movement to provide a theoretical backdrop to the concept of CoPP.

Insights from counter-hegemony and social movements

Hegemony translates as 'dominance over', originally a system of class alliance in which the dominant class tried to win over subordinate classes (Gramsci, 1971). Since then, the concept of hegemony has been applied to various forms of dominance, such as global hegemony in international relations (Cox, 1981); hegemony of masculinity (Connell & Messerschmidt, 2013; Carrigan, Connell, & Lee, 1985); environmental politics (Falkner, 2005); or racial hegemony (Winant, 2006).

In the context of gender inequality, hegemony has been applied as an analytical lens to meritocracy and individualism ideologies in HE, and sexual harassment discourse (Seron et al., 2018; McDonald & Charlesworth, 2013); and male hegemonic whiteness in HE (Cabrera, 2016). These applications have illuminated mechanisms of cultural reproduction of gender inequality by both men and women, such as media misrepresentations or limited opportunities for awareness-raising, and gendered meritocracy as 'common sense' (Ramos, 1982).

In response to hegemony, an alternative position of values and principles can emerge in the form of a social movement or group – counter-hegemony – to challenge or supplant the dominant discourse. This can be achieved through coalescing social forces that reject the hegemonic 'common sense' and assert a new, revised world view promoting the emancipation of the subjugated, or at least challenging the status quo. New social movements in Western Europe looked beyond class conflict and united around other issues such as race, environment, politics, homosexuality, and feminism (e.g., Black Lives Matter; Extinction Rebellion, #metoo, the Women Strike, the Yellow Vests, etc.). Collective action for change was rooted in desire for legitimation of different identities and lifestyles, and changes to hegemonic normative and cultural codes (Polletta & Jasper, 2001).

Furthermore, studies of gender and sexuality have challenged dominant theoretical approaches that narrowly define social movements through masculinist assumptions of who is an activist, and what counts as activism (Wulff, Bernstein, & Taylor, 2015). Gender and sexuality movements, for instance, that seek material and symbolic change in institutions have been traditionally dismissed by political process theories as 'expressive', rather than 'legitimate' forms of activism (Bernstein & Olsen, 2009; Gilmore & Kaminski, 2007). However, women's and LGBT movement studies have illuminated that the strategic use of identity influences movement goals, outcomes, tactics, and targets. Therefore, social movements can target a variety of state and non-state institutions, since domination in modern society results from multi-layered sources of power, not merely the state or economy. Examples of such power sources include practices, institutions, cultural norms, knowledge systems, or organisations (Acker, 1990) in a variety of contexts creating idiosyncratic conditions that stimulate particular forms of mobilisation (Wulff et al., 2015).

The American women's suffrage movement – an example of a social movement community (Buechler, 1990) – illustrates how a network of women's rights activists sustained and mobilised itself over time despite the absence of formal organizations (cited in Hassan & Staggenborg, 2015). As numerous feminist groups had varying degrees of self-sustainment, a women's movement maintained itself in communities where women were present. This was amplified, for example, in women's centres, women's studies programmes at universities, feminist bookstores and social venues, women's health centres, and rape crisis centres, among others (Buechler, 1990 cited in Hassan & Staggenborg, 2015). It was the cultural, institutional, and political elements of the women's movement community that have maintained feminism in the United States (Staggenborg, 1998), but also an interplay of individual and collective identity (Stoecker,1995).

To supplement conceptualising social movements as simply 'reactive', new ways of thinking emerged about the 'processes' by which the resources necessary for collective action are mobilised (Della Porta & Diani, 2020). Tilly (1978) defined a social movement as a rational, purposeful, and organised action which takes into consideration available resources, calculates the costs and benefits, strategically organises itself, and interacts to spear its development (Della Porta & Diani, 2020). Therefore, social movements are more than anonymous, loose, spontaneous, and organic, and can be intentional with strong organising and rational processes, similar to how CoPs can be developed (Wenger et al., 2002) if their domain and practice aim for institutional (here: political) change.

Communities of Political Practice

Chapter 1 of this volume elaborates on the classical CoP theory, which we do not wish to replicate. Thus briefly, the structural model of CoP is based around three key elements always present at the same time: *domain*, *practice*, and *community*, which make up an 'ideal knowledge structure' (Wenger et al., 2002). CoPs create common ground and identity through a well-defined *domain*, which gives the CoP its *raison d'être* and provides potential value to the community members and other peripheral stakeholders. The domain provides an inspiration for the members to engage, and contribute, and makes their actions and learning meaningful. The *community* is the 'social fabric of learning' (Wenger et al., 2002, p. 28) and promotes interactions and relationships between people. Trust and mutual respect are needed for learning, sharing experiences, ideas, knowledge, and exposing one's vulnerabilities. Lastly, whilst the *domain* is the topic of the CoP, the *practice* denotes the specific stocks of knowledge the community works with. It is a rich combination of norms, tacit knowledge, frameworks, ideas, tools, information, styles, language, stories, case studies, and documents.

CoP is a 'learning partnership among people who find it useful to learn from and with each other [and] use each other's experience of practice as a learning resource' (Wenger et al., 2011, p. 9). Thus, CoP members can approach institutional change iteratively by assessing what works in their organisation and sharing progress, successes and potential pitfalls within the community, thus contributing to the partnership of mutual learning, practice exchange, and building of new knowledge. Furthermore, gender equality projects in HE and R&I require *feminist* knowledge transfer based on self-reflexive endeavour, 'safe exploration of new ideas and expanding the boundaries of knowledge from multiple experiences' (Bustelo, Ferguson, & Forest, 2016, p. 37). Community members linked through a dyadic relationship between more and less experienced centres or entities can facilitate learning and sharing best practices (Wenger, 1998) focused on change implementation plans. But CoPs could also initiate change from the ground by collaborating and engaging in innovative practice and scaling it up so that this alternative practice can then influence and become part of the wider structure. Radical innovations frequently occur at the interstices across communities (Swan, Scarbrough, & Robertson, 2002).

As Bridwell-Mitchell (2016) contends institutional change can be fostered through ad hoc peer collaboration rather than political contestation. Indeed, the Ferguson report on structural change for gender equality in R&I recognises that challenges arise from the integration of 'previously inactive institutions and countries in structural change. This may require approaches that are adapted to local and national gender equality contexts, as well as linkages with more advanced and experienced actors' (Ferguson, 2021, p. 40), suggesting an alignment between such endeavours and a CoP approach.

CoP theory has not escaped critique, for instance, questioning whether analyses of CoPs should presume the pre-existence of an objective, socio-historical contexts and actors, or whether they should be sought through tracing nests of practice that would point to their existence. In other words, should we speak of Communities of Practice, or practice of communities (PoCs) (Gherardi, 2009)? This is pertinent in conceptualising CoPPs, because practice and procedures are what sustains a hegemonic order in institutions (Conran & Thelen, 2016) and practices reproduce structures (Giddens, 1984). Thus, it is an alternative practice that can change the status quo. Gherardi (2009) de-emphasises community and argues for viewing the community as an effect as well as a device for the reproduction of the field of practices, awarding practice the prominence through the articulation of PoC, rather than CoP. This conceptualisation fits well when broadly considering academia or research as a 'practice', defined as an institutionalised body of knowledge (Gherardi, 2009). However, as in the case of ACT CoPs, what brings people together (Latour, 2007) is an issue, or a certain situation, a collective objection to the status quo as evidenced by objective gender statistics and subjective perceptions of women in HE and R&I demonstrating inequality and discrimination which then leads to the practice of gender equality work. As such, it is the matter of concern (or *domain*) that should be afforded equal footing, as it triggers the innovation of alternative practice (which is necessarily political) and glues the community together. The uptake and the transformative promise of alternative practice are contingent on the appetite of the community to adopt it, and its enactment by purposeful and cognisant individuals converged in this community, rather than puppet-like actors in a vacuum. Alternative practice needs agency to be produced and to challenge status quo, and at the same time, agency needs alternative practice to change structures (Stones, 2005).

Institutional change through interactions, innovative or alternative practices and embedded actions is a political endeavour (Drazin, 1990). Consequently, 'changing practice often becomes controversial since it implies a redefinition of its configuration, power relations and ways of knowing' (Mørk et al., 2010, p. 576). As Latour (2007) observed, politics is issue-oriented as it turns around matters of concern, and it brings people together *because* they disagree. To understand how CoP could be designed for institutional change, social movement theory illuminates crucial elements for CoPs if the domain is counter-hegemonic, and the practice politically transformative. Further, Buechler (1990) argued that some aspects of social movements are better understood as communities since social movement communities include *both* organisations and informal networks of activists, cultural networks, institutional supporters, alternative institutions, as well as other actors who share a passion for the goals of a social movement (Staggenborg, 1998).

As such, we draw on Gläser's (2004, p. 7) thinking on social movement as 'a community whose identity is based on the perception of a common goal [...] – namely the intention to bring about social change'. Social movements have three characteristics: members hold opposing orientations to clearly identified opponents; they connect through dense informal networks; and they share a distinct collective identity (Della Porta & Diani, 2020). Community membership relies on an enduring identification with a particular community. The notion of collective identity within social movements explicitly moves beyond personal identity, and collective identity is defined as 'an individual's cognitive, moral, and emotional connection with a broader community, category, practice, or institution [and] is a perception of a shared status or relation' (Polletta & Jasper, 2001, p. 285). Collective identity is constructed on an ongoing basis in members' interactions, and the development of a social movement is contingent on successful creation of collective identity in the first place (Melucci, 1989). The relationship is reinforced further because collective identity 'translates structural inequality and injustice into individual dissatisfaction, [and] is a primary social-psychological dynamic of mobilisation' (Wulff et al., 2015, p. 110). The relationship is also political, as it brings together people that disagree about a shared matter of concern (Latour, 2007).

CoPs have a unique opportunity to promote change through learning, sharing knowledge, and importantly enacting alternative practices, which social movements may lack. But like social movements, communities need a structured way of *organising* (Gläser, 2004; Tilly, 1978). Collective action, but also arguably collaborative practice, demands coordination, continuity, actors' capacity to identify themselves as members of a given community, leadership, and support (Billings, 1990); and historically, these functions have been performed by *organisations* (Della Porta & Diani, 2020), enabling social movements to take advantage of political opportunities to influence policy outcomes (Hafner-Burton & Pollack, 2002).

The work of Bridwell-Mitchell (2016) also illuminates viability of institutional change through collaborative institutional agency, built on cohesive communities with moderate demographic diversity to maximise potential for change. As a result, innovation is seeded by trusted peers and reinforced through strong socialisation pressures, leading to cognitive and normative convergence in members' understandings, aims, and transformative practices. This is particularly significant for inter-organisational, transnational CoPs, whose membership might comprise people representing multiple and diverse organisations.

As in many social entities, both social movements and communities have members in power based on merit drawn from superior knowledge pertinent to the domain and practice, a track record of effective past actions and key relationships. This is classified by Gläser (2004) as *stratification,* which creates an elite that holds positions in the formal or informal organisations within social movements and communities. For example, in the GenderTime EC project, 'transfer agents' were assigned to act outside the CoP to transfer and implement gender equality knowledge in a strategic manner (Thaler, 2016). Involving individuals in powerful and relevant positions committed to the idea of gender equality in R&I was necessary to secure wider support for the implementation of gender equality plans. This stratification is also linked to CoP members overlapping membership across many communities (i.e., HE senior management and CoP) (Thaler, 2016). This is the third commonality between a social movement and a community: the possibility for the members to belong to more than one community or social movement, or *membership overlap* (Gläser, 2004). This can produce effective synergies (but also challenging contradictions) across communities. For CoP, this would mean that effective practice can be cross-pollinated from other domains, thus further enriching and enabling transformative capacities of CoP.

Organising for institutional change:
ACT Communities of Practice

To foster social or institutional change, a community needs to organise itself to be effective (Della Porta & Diani, 2020; Gläser, 2004; Hafner-Burton & Pollack, 2002; Billings, 1990; Tilly, 1978). To examine ACT CoPs in relation to ways of organising, we draw on the Ahrne and Brunsson's (2010) definition of organisation as a 'decided order' based on five elements of organisation: membership, rules, hierarchy, monitoring, and sanctioning.

The eight ACT CoPs operating under the ACT-on-Gender project were co-created, but there was also a degree of organic emergence and choice of priorities. At the beginning of the project, a survey was conducted to understand the extent of existing practices, needs and networks around gender equality in research performing and funding organisations across Europe and beyond. More importantly, the survey was to identify potential members of ACT CoPs. Social network analysis showed existing cooperation clusters and identified both central and disconnected actors (Reidl et al., 2019). The processes leading to decisions regarding the choice of certain members and the definition of criteria for membership were intensive. The focus was largely on the structure of interorganisational networks, which started from evaluation of their density and connectedness (Diani, 2015), and also who could benefit most. Institutions that had previous experience of implementing institutional change as well as some institutions that had various international networks (through belonging to structural change projects) reached out to previous contacts and made new contacts identified through a community mapping survey. As a result, CoP members were clustered into eight thematic, discipline, or geographic CoPs that included varying levels of experience in GEP implementation and institutional change to ensure the less experienced institutions learned from the more experienced.

Membership was voluntary (individual, not institutional) as long as it was affiliated to a research funding or performing organisation. Members included both practitioners and researchers. As Wenger et al. (2002, p. 36) suggest, CoPs do not have to be purely spontaneous and its membership entirely voluntary, but the CoP's success will depend on participation and personal investment of the members. Therefore, the CoP's collective political motivation of members was key. Membership overlap, that is belonging to more than one community simultaneously was part and parcel of participating in the ACT CoP. Members were scientists and academics who also worked or engaged with other circles in different fields from past or consecutive gender equality projects and initiatives.

The *rules* of engagement imposed on the subscribing CoPs were included in the memorandum of understanding signed between the members and the project coordinator and were not legally binding. It specified good faith commitments of the members. These written rules were supplemented with more informal norms developed within the communities via idiosyncratic processes of socialisation and internalisation during the lifetime of the project (Ahrne & Brunsson, 2010), such as meeting attendance and engagement.

Each CoP did not rely on a *hierarchy* as such, however, the eight CoPs were theoretically subjected to decision-making processes influenced by the steering committee and the project coordinator who was, in turn, answerable to the EC project officer. Each CoP had a dedicated facilitator to broker relationships and connect people responsible for organising and supporting the community through planned meetings, agenda-setting, workshops, webinars, blogs, e-discussions, access to resources and gender experts; and acted as a link between the CoP and the project consortium. However, the role of the CoP facilitator was not to create a hierarchical channel to navigate through, but rather to foster horizontal relationships (Wenger et al., 2002, p. 128).

The project coordinator carried out formal and informal *monitoring* of the CoPs' activities and outcomes of collaboration as part of project-level evaluations, rather than monitoring the CoPs themselves. Both qualitative and quantitative questionnaires were conducted to gauge the level of member collaboration and engagement with the CoPs' aims for the purposes of project deliverables, but also informal monitoring occurred during online project consortium meetings where updates and progress of the CoPs were reported by each facilitator. The possibility of *sanctions* was also embedded in the memorandum of understanding, such that participation in the CoP is terminated if the member no longer fulfilled the criteria for participation or failed to meet its obligations, or acted contrary to the aims and objectives or values of the project.

Within the CoPs, there was no internal *stratification* in place as such, however the external support provided by the ACT project consortium consisted of an experienced academic and professional community expert in gender equality and GEP implementation knowledge, past successful outputs and projects within the domain of gender equality, and professional relations with other members of the community (Gläser, 2004). Members of these groups held positions in the formal organisations of HE and R&I who formed part of the consortium. The consortium consisted of core partners responsible for creating tools and designing support mechanisms for the CoPs, and seed partners directly involved in setting up, supporting the CoPs and recruiting and managing CoP facilitators. The advisory board provided the breadth and depth of collective expertise in human resource strategies and practices in research institutions and science programmes and related research. The only deliberate stratification inside CoPs related to the level of expertise and experience in gender equality progress within institutions to enable mutual learning.

The ACT CoPs were able to leverage the expertise and support of these actors at any point. There were three types of *support* with varying levels of intensity available to the ACT CoPs. This included CoP facilitation, collaboration events, workshops and conference organisation, access to knowledge and best practices resources, financial support, and self-managed online support. Cumulatively, this membership allowed participants to receive cross-mentoring in a peer support network and equipped them as change agents regardless of their hierarchical position in the institution (Leenders, Bleijenbergh, van den Brink, 2020).

As Lave and Wenger (2001) argued identity is inseparable from issues of practice, community, and meaning. The *collective identity* uniting the ACT

CoP members was extant in learning and sharing the practice and technicalities of GEP implementation, institutional change, gender knowledge, and strategies against resistance to gender equality. However, we also noted that most of the ACT CoP members were female, as understandably women are more likely than men to be aware of the negative side of the stereotypical and gendered nature of institutions in which they work (Acker, 2000). Thus, they are also more likely to actively engage; but this can be seen as a significant caveat in gender equality projects, as the participating women are casting themselves as victims fighting for their own rights with token male allies. They are also disproportionately burdened by responsibilities for raising awareness of and tackling discrimination (Tzanakou, 2019). Moreover, social change toward gender equality cannot be achieved without the active support and participation of men (Schacht & Ewing, 2004).

On the other hand, as Bernstein (2005) puts it, personal subjectivity derived from experiences of oppression can be empowering and provide the motivation for social change. A collective interpretation and analysis of material conditions, the lived experience, and social location of participants may facilitate translating personal experience into collective action. Thus, CoP female members were not just a 'group in itself', united only by the shared status of being 'women who work in HE or R&I', but they had the potential to become a 'group for itself' (Górska & Bilewicz, 2015) through mutual recognition of shared exploitative experience, regular interactions, and finding a safe space to find their voice. Collective construal of the academic female members' disadvantage and perceiving deprivation at group-level rather than individual-level influence how people engage in social change (Górska & Bilewicz, 2015). Such collective identity in designed, intentional CoPs is strengthened through the counter-hegemonic domain of gender equality, which can provide a cognitive, moral, and emotional connection with a broader community to foster active participation in institutional change (Polletta & Jasper, 2001). Even after the project ends, and the CoPs face the threat of losing the formal support if no further funding opportunities arise, the existence of collective identity linking members to each other will enable them to feel part of the same collective effort in the future and develop further joint actions (Diani, 2012) (e.g., CoP partially reconfiguring as a COST Action, or other CoPs co-designing new grant proposals for Horizon Europe).

Conclusion

Institutional change does not have to be necessarily grounded in political contestations, but it can be fostered through peer collaboration and alternative practice. Indeed, 'often, change happens through the "institutional work" of agents who overturn deeply ingrained, taken-for-granted, value-laden practices' (Bridwell-Mitchell, 2016, p. 161). Thus, the purpose of this chapter was to argue and demonstrate that CoPs might be a particularly powerful method of organising equality projects and institutional change. To support this claim, we drew on theories of counter-hegemony, social movement, and CoP and proposed a new concept of *CoPP*, defined as a group of institutionally affiliated people across different organisations or nations coalescing around a shared concern for social equality who engage in transformative practice, who learn from each other and co-create knowledge and other useful resources through regular interactions to act on institutional change. In so doing, we moved beyond the scholarly focus on situated and peer learning of CoP.

Instead, we have illuminated how CoPPs can contribute to inequality contestations in institutions with skewed distributions of power and resources, with the benefit of being institutionally embedded and understanding the idiosyncrasies of the context. We have proposed that CoPP is characterised by its counter-hegemonic *domain*, transformative, alternative (and political) *practice*, and a purposefully *organised* and *supported community* that share a *collective identity* and passion for achieving equality. Looking closely into the inner workings of the externally funded ACT-on-Gender CoP project, we have demonstrated how CoPPs can be intentionally established considering their organisational elements of membership, member overlaps, hierarchy and stratification, rules, monitoring and sanctions, and the targeted economic and non-economic support provided by project infrastructures.

Future research should pay attention to the dynamics of practice leading to collective action and eventually institutional change. The limitation is that we have mainly focused on the structural conditions for action. However, what is still unexplored is the intricacies of how institutional change happens when it is fostered by CoPPs, and how its actors enact the transformative practice they have so astutely learnt, shared, and co-created through CoPPs' mutual engagement. The question remains: to what extent is this form of organising for change transformative or reproductive of institutional structures and inequality regimes? Nevertheless, reconceptualising CoP as CoPP creates an opportunity for practitioners to consider this way of organising for change projects. It also provides a fertile ground for scholars to further CoP theory and illuminate the crossroads of CoP and social movement theories, fleshing out potential threats and opportunities if a CoPP is to be institutionally transformative.

References

Acker, Joan (1990). Hierarchies, Jobs, Bodies: A Theory of Gendered Organizations, *Gender & Society*, 4(2), 139–158. doi.org/10.1177/089124390004002002

Acker, Joan (1992). Gendered Institutions – From Sex Roles to Gendered Institutions. *Contemporary Sociology-a Journal of Reviews*, 21(5), 565–69. doi.org/10.2307/2075528

Acker, Joan (1998). The Future of 'Gender and Organizations': Connections and Boundaries. *Gender, Work & Organization*, 5(4), 195–206. doi.org/10.1111/1468-0432.00057

Acker, Joan (2000). Gendered Contradictions in Organizational Equity Projects [Editorial Material]. *Organization*, 7(4), 625–32. doi.org/10.1177/135050840074007

Ahrne, Göran, & Brunsson, Nils (2010). Organization Outside Organizations: The Significance of Partial Organization. *Organization*, 18(1), 83–104. doi.org/10.1177/1350508410376256

Albenga, Viviane (2016). Between Knowledge and Power: Triggering Structural Change for Gender Equality from Inside in Higher Education Institutions. In (Eds.), Maria Bustelo, Lucy Ferguson & Maxime Forest. *The Politics of Feminist Knowledge Transfer: Gender Training and Gender Expertise*, Palgrave McMillan, Basingstoke.

Amin, Ash, & Roberts, Joanne (2008). Knowing in Action: Beyond Communities of Practice. *Research Policy*, 37(2), 353–69.

Annala, Johanna, & Mäkinen, Marita (2017). Communities of Practice in Higher Education: Contradictory Narratives of a University-Wide Curriculum Reform. *Studies in Higher Education*, 42(11), 1941–57.

Barnard, Sarah (2017). The Athena SWAN Charter: Promoting Commitment to Gender Equality in Higher Education Institutions in the United Kingdom. In White, Kate & O'Connor, Pat (Eds.). *Gendered Success in Higher Education*. Palgrave Macmillan, London. doi.org/10.1057/978-1-137-56659-1_8

Bernstein, Mary (2005). Identity Politics. In *Annual Review of Sociology. Annual Reviews*, 31, 47–74. doi.org/10.1146/annurev.soc.29.010202.100054

Bernstein, Mary and Olsen, Kristine A. (2009). Identity Deployment and Social Change: Understanding Identity as a Social Movement and Organizational Strategy. *Sociology Compass*, 3, 871–83.

Billings, Dwight B. (1990). Religion as Opposition: A Gramscian Analysis. *American Journal of Sociology*, 96(1), 1–31.

Bridwell-Mitchell, Ebony N. (2016). Collaborative Institutional Agency: How Peer Learning in Communities of Practice Enables and Inhibits Micro-Institutional Change. *Organization Studies*, 37(2), 161–192. doi.org/10.1177/0170840615593589

Brown, John Seely, & Duguid, Paul (1991). Organizational Learning and Communities-of-Practice: Toward a Unified View of Working, Learning, and Innovation. *Organization Science*, 2(1), 40–57.

Buechler, Steven M. (1990). *Women's Movements in the United States*. Rutgers University Press, New Brunswick, NJ.

Bustelo, María, Ferguson, Lucy & Forest, Maxime (2016). *The Politics of Feminist Knowledge: Gender Training and Gender Expertise*. Palgrave Macmillan UK.

Cabrera, Nolan L. (2016). When Racism and Masculinity Collide: Some Methodological Considerations from a Man of Colour Studying Whiteness. *Whiteness and Education*, 1(1), 15–25. doi.org/10.1080/13613324.2015.1122662

Carrigan, Tim, Connell, Bob, & Lee, John (1985). Toward a New Sociology of Masculinity. *Theory and Society*, 14(5), 551–604.

Carroll, William K., & Ratner, R. S. (1994). Between Leninism and Radical Pluralism: Gramscian Reflections on Counter-Hegemony and the New Social Movements. *Critical Sociology* 20(2), 3–26.

Childs, Sarah (2004). A Feminised Style of Politics? Women MPs in the House of Commons. *British Journal of Politics and International Relations* 6(1), 3–19.

Connell, Robert W., & Messerschmidt, James W. (2013). Hegemonic Masculinity: Rethinking the Concept. *Revista Estudios Feministas*, 21(1), 241–82. doi.org/10.1590/S0104-026X2013000100014

Conran, James, & Thelen, Kathleen (2016). Institutional Change. In Fioretos. Orfeo, Tulia, Falleti & Adam Sheingate (Eds.), *The Oxford Handbook of Historical Institutionalism*. Oxford University Press, Oxford. doi.org/10.1093/oxfordhb/9780199662814.001.0001

Cotterill, Pamela, Jackson, Sue, & Letherby, Gayle (2007). Challenges and Negotiations for Women in Higher Education. Springer, The Netherlands. doi.org/10.1007/978-1-4020-6110-3

Cox, Robert W. (1981). Social Forces, States and World Orders: Beyond International Relations Theory. *Millennium*, 10(2), 126–55. doi.org/10.1177/03058298810100020501

Dahlerup, Drude (1988). From a Small to a Large Minority: Women in Scandinavian Politics. *Scandinavian Political Studies*, 11(4), 275–98.

Della Porta, Donatella, & Diani, Mario (2020). *Social Movements: An Introduction* (3rd ed.). Wiley Blackwell. New Jersey.

Diani, Mario (2012). Attributes, Relations, or Both? Exploring the Relational Side of Collective Action. *Contributions to Nepalese Studies*, 39(Special Issue), 21–44.

Diani, Mario (2015). *The Cement of Civil Society*, Cambridge University Press, New York.

DiMaggio, Paul. J., & Powell, Walter W. (1995). *The New Institutionalism in Organizational Analysis* (2nd ed.). University of Chicago Press, Chicago.

Drazin, Robert. (1990). Professionals and Innovation: Structural-Functional versus Radical-Structural Perspectives. *Journal of Management Studies*, 27(3), 245–63.

European Commission. (2012). *A Reinforced European Research Area Partnership for Excellence and Growth*. European Commission. Retrieved from https://eur-lex.europa.eu/LexUriServ/LexUriServ.do?uri=COM:2012:0392:FIN:EN:PDF [08.09.2021].

European Commission, Directorate-General for Research and Innovation, (2019). *SHE Figures 2018*. Publications Office. Retrieved from https://data.europa.eu/doi/10.2777/936 [19.05.2022].

Eurostat. (2019). Teaching Staff in Tertiary Education by Sex and Level of Education, 2017. Retrieved from https://ec.europa.eu/eurostat/statistics-explained/index.php?title=File:Teaching_staff_in_tertiary_education_by_sex_and_level_of_education,_2017_(thousands)_ET19.png#filehistory [07.07.2020].

Falkner, Robert (2005). American Hegemony and the Global Environment. *International Studies Review*, 7(4), 585–99.

Ferguson, Lucy (2021). *Analytical Review: Structural Change for Gender Equality in Research and Innovation*. Publications of the Ministry of Education and Culture, Finland, 2021:4, Helsinki.

Gherardi, Silvia (2009). Community of Practice or Practices of a Community. In S. J. Armstrong & C. V. Fukami (Eds.), *The Sage Handbook of Management Learning, Education, and Development* (pp. 514–530). SAGE Publications Ltd, London, Thousand Oaks, New Delhi, Singapore.

Giddens, Anthony (1984). *The Constitution of Society*, Polity Press, Cambridge.

Gilmore, Stephanie and Kaminski, Elizabeth (2007). A Part and Apart: Lesbian and Straight Feminist Activists Negotiate Identity in a Second-Wave Organization. *Journal of the History of Sexuality.* 16(1), 95–113.

Gläser, Johann (2004). Social Movements as Communities. In K. Richmond (Ed.), *TASA Annual Conference, "Revisioning Institutions: Change in the 21st Century", Refereed Conference Papers.* (pp. 1–10): The Australian Sociological Association (TASA).

Goodyear, Victoria, Casey, Ashley, & Kirk, David (2016). Practice Architectures and Sustainable Curriculum Renewal. *Journal of Curriculum Studies*, 49(2), doi. org/10.1080/00220272.2016.1149223

Górska, Paulina, & Bilewicz, Michał (2015). When "a Group in Itself" Becomes "a Group for Itself": Overcoming Inhibitory Effects of Superordinate Categorization on LGBTQ Individuals. *Journal of Social Issues*, 71(3), 554–75. doi.org/10. 1111/josi.12128

Gramsci, Antonio (1971). *Selections from the Prison Notebooks of Antonio Gramsci.* International Pubs, London.

Gramsci, Antonio (1985). *Selections from Cultural Writings.* Harvard University Press, London.

Grey, Sandra (2006). Numbers and Beyond: The Relevance of Critical Mass in Gender Research. *Politics & Gender*, 2, 491–530.

Hafner-Burton, Emilie, & Pollack, Mark A. (2002). Mainstreaming Gender in Global Governance. *European Journal of International Relations*, 8(3), 339–73. doi.org/10.1093/oxfordhb/9780199678402.013.37

Hassan, Hatem M. & Staggenborg, Suzanne (2015). Movements as Communities. In Donatella Della Porta & Mario Diani (Eds.) The Oxford Handbook of Social Movements. Oxford University Press, Oxford. doi.org/10.1093/oxfordhb/ 9780199678402.013.59

Higher Education Statistics Agency. (2020). *Who's Working in HE? Personal Characteristics.* Higher Education Statistics Agency. Retrieved from https://www.hesa. ac.uk/data-and-analysis/staff/working-in-he/characteristics [07.07.2020].

Hutchins, Brett, & Boyle, Raymond (2017). A Community of Practice Sport Journalism, Mobile Media and Institutional Change. *Digital Journalism*, 5(5), 496–512. doi.org/10.1080/21670811.2016.1234147

Iaquinto, Ben, Ison, Ray, & Faggian, Robert (2011). Creating Communities of Practice: Scoping Purposeful Design. *Journal of Knowledge Management*, 15(1), 4–21. doi.org/10.1108/13673271111108666

Kendall, Kathleen, Collett, Tracey, de Iongh, Anya, Forrest, Simon, & Kelly, Moira (2018). Teaching Sociology to Undergraduate Medical Students. *Medical Teacher*, 40(12), 1201–7. doi.org/10.1080/0142159X.2018.1505038

Lave, Jean. & Wenger, Etienne, (2001). Legitimate Peripheral Participation in Communities of Practice, in (Eds.) Julia Clarke, Ann Hanson, Roger Harrison & Fiona, Reeve, *Supporting Lifelong Learning*, Routledge, London. doi.org/10. 4324/9780203996287

Leenders, Joke, Bleijenbergh, Inge, & Van den Brink, Marieke C. L. (2020). *Myriad Potential for Mentoring: Understanding the Process of Transformational Change through a Gender Equality Intervention*, 27, 379–94. doi.org/10.1111/gwao. 12385

Latour, Bruno (2007). Turning around Politics: A Note on Gerard de Vries' Paper. *Social Studies of Science*, 37(5), 811–20. doi.org/10.1177/0306312707081222

Lombardo, Emanuela, & Meier, Petra (2006). Gender Mainstreaming in the EU: Incorporating a Feminist Reading? *European Journal of Women's Studies*, 13(2), 151–66. doi.org/10.1177/1350506806062753

Lovenduski, Joni (2001). Women and Politics: Minority Representation or Critical Mass? *Parliamentary Affairs*, 54(4), 743–58.

McDonald, Paula & Charlesworth Sara (2013). Framing Sexual Harassment through Media Representations. *Women's Studies International Forum*, 37, 95–103.

McIntyre, Fiona (2020). Athena SWAN Awards Axed from NIHR Funding Requirement. Research Professional News. Retrieved from https://www. researchprofessionalnews.com/rr-news-uk-charities-and-societies-2020-9-athena-swan-awards-axed-from-nihr-funding-requirement/ [08.09.2021]

Mahoney, James, & Thelen, Kathleen (2010). A Theory of Gradual Institutional Change. *Explaining Institutional Change*, 1–37. doi.org/10.1017/CBO9780511806414. 003

Melucci, Alberto. (1989). *Nomads of the Present: Social Movements and Individual Needs in Contemporary Society*. Hutchinson Radius, London.

Mørk Bjørn, Erik, Hoholm, Thomas, Ellingsen, Gunnar, Edwin, Bjørn, & Aanestad, Margunn (2010). Challenging Expertise: On Power Relations within and across Communities of Practice in Medical Innovation. *Management Learning*, 41(5), 575–92. doi.org/10.1177/1350507610374552

O'Hagan, Angela, Klatzer, Elisabeth (Eds.). (2018). *Gender Budgeting in Europe: Developments and challenges*. Palgrave Macmillan, Cambridge.

Polletta, Francesca, & Jasper, James M. (2001). Collective Identity and Social Movements. *Annual Review of Sociology*, 27(1), 283–305. doi.org/10.1146/annurev. soc.27.1.283

Ramos, Valeriano (1982). The Concepts of Ideology, Hegemony, and Organic Intellectuals in Gramsci's Marxism. *Theoretical Review*, 27 (Mar-Apr). Retrieved from https://www.marxists.org/history/erol/periodicals/theoretical-review/1982301. htm [05.05.2021].

Reidl, Sybile, Krzaklewska, Ewa, Schön, Lisa, & Warat, Marta (2019). ACT Community Mapping Report: Cooperation, Barriers and Progress in Advancing Gender Equality in Research Organisations. Retrieved from https://zenodo.org/record/3247433#.YJKX7bVKiUk [05.05.2021].

Schacht, Steven P., & Ewing, Doris W. (2004). *Feminism with Men: Bridging the Gender Gap*. Rowman & Littlefield, Lanham, Md.

Seron, Carroll, Silbey, Susan, Cech, Erin, & Rubineau, Brian (2018). "I am Not a Feminist, but...": Hegemony of a Meritocratic Ideology and the Limits of Critique among Women in Engineering. *Work and Occupations*, 45(2), 131–67. doi.org/10.1177/0730888418759774

Staggenborg, Suzanne (1998). Social Movement Communities and Cycles of Protest: The Emergence and Maintenance of a Local Women's Movement. *Social Problems*, 45(2), 180–204.

Stoecker, Randy (1995). Community, Movement, Organization: The Problem of Identity Convergence in Collective Action. *The Sociological Quarterly*, 36(1), 111–30.

Stones, Rob (2005). *Structuration Theory*. Palgrave Macmillan, Houndmills, Basingstoke, Hampshire; New York.

Streeck, Wolfgang & Thelen, Kathleen Ann (Eds.). (2005). *Beyond Continuity Institutional Change in Advanced Political Economies*. Oxford University Press; New York.

Swan, Jack, Scarbrough, Harry, & Robertson, Maxine (2002). The Construction of 'Communities of Practice' in the Management of Innovation. *Management Learning*, 33(4), 477–96. doi.org/10.1177/1350507602334005

Teelken, Christine, & Deem, Rosemary (2013). All are Equal, but Some are More Equal than Others: Managerialism and Gender Equality in Higher Education in Comparative Perspective. *Comparative Education*, 49(4), 520–35. doi.org/10.1080/03050068.2013.807642

Thaler, Anita (2016). *Learning Organisations in Science and Research: The Role of Transfer Agents in Gender Equality Change Processes.* Retrieved from https:// www.researchgate.net/publication/301540864_Learning_Organisations_in_ Science_and_Research_The_Role_of_Transfer_Agents_in_Gender_Equality_ Change_Processes [05.05.2021].

Tilly, Charles (1978). *From Mobilization to Revolution.* McGraw-Hill, Reading.

Tzanakou, Charikleia (2019). Unintended Consequences of Gender-Equality Plans. *Nature,* 570 (277). doi.org/10.1038/d41586-019-01904-1

Tzanakou, Charikleia & Pearce, Ruth (2019). Moderate Feminism within or against the Neoliberal University? The Example of Athena SWAN. *Gender Work & Organization,* 26, 1191–211. doi.org/10.1111/gwao.12336

Verge, Tània, Mariona Ferrer-Fons, & M José González. (2018). Resistance to Mainstreaming Gender into the Higher Education Curriculum. *European Journal of Women's Studies,* 25(1) 86–101. doi.org/10.1177/1350506816688237

Waylen, Georgina (2014). Informal Institutions, Institutional Change, and Gender Equality. *Political Research Quarterly,* 67(1), 212–23. doi.org/10.1177/1065912913510360

Wenger, Etienne (1998). *Communities of Practice: Learning, Meaning, and Identity.* Cambridge University Press, Cambridge.

Wenger, Etienne, McDermott, Richard A., & Snyder, William (2002). *Cultivating Communities of Practice: A Guide to Managing Knowledge.* Harvard Business School Press, Boston.

Wenger, Etienne, Trayner, Beverly, & de Laat, Maarten (2011). Promoting and Assessing Value Creation in Communities and Networks: A Conceptual Framework. Retrieved from https://wenger-trayner.com/resources/publications/evaluation-framework/ [05.05.2021].

Winant, Howard (2006). Race and Racism: Towards a Global Future. *Ethnic and Racial Studies,* 29(5), 986–1003. doi.org/10.1080/01419870600814031

Wulff, Stephen, Bernstein, Mary & Taylor, Verta (2015). New Theoretical Directions from the Study of Gender and Sexuality Movements: Collective Identity, Multi-Institutional Politics, and Emotions. In Donatella Della Porta & Mario Diani (Eds.) *The Oxford Handbook of Social Movements.* Oxford University Press, Oxford. doi.org/10.1093/oxfordhb/9780199678402.013.59

3 The development of the Gender Equality Audit and Monitoring survey

Kevin Guyan, Amanda Aldercotte, Jörg Müller, Maria Caprile and Sergi Yanes Torrado

Introduction

To address gender inequality within higher education and research organisations, it is first vital to define the problem. In place of something abstract and fluid, gender equality practitioners need to establish the scale and contours of the challenge they face – attribute a name to the challenge and describe it with characteristics fixed in time and space. As described by Rachel Palmén and Jörg Müller in the Introduction chapter, "knowing the institution" is an integral component in the ability of Communities of Practice (CoPs) to develop strategies and initiatives that respond to gender inequality. An approach that builds on the European Commission's (2011) definition of structural change, alongside the requirement for top-level support and effective management practices. This early step in the process is facilitated by the Gender Equality Audit and Monitoring (GEAM) survey, one of the main outputs of the ACT project.[1] Developed by research teams from the United Kingdom (Advance HE) and Spain (Notus and Universitat Oberta de Catalunya), the GEAM provides a comprehensive, transferable and transnational survey for higher education and research organisations that wish to monitor and/or undertake an audit of gender equality of academic, technical and support staff. The modular survey includes units on working conditions, beliefs and biases, organisational culture and climate and individual behaviours as well as the capture of data about the identity characteristics of respondents.[2] This chapter reflects on decisions made in the design of the GEAM, its roll-out and its revision. As well as charting the development of the GEAM, it also presents a case study that provides an in-depth consideration of its approach to the collection of gender, sex and sexuality data.

The GEAM survey consists of core and extended questions. Core questions comprise a comprehensive collection of questions that cover most aspects of gender equality in research organisations and provides a good starting point for implementing an initial audit and assessment of an organisation or

DOI: 10.4324/9781003225546-3

organisational unit's gender equality. The core survey is also fixed in that the questions are set for all organisations using it, and are thus directly comparable across these organisations. Extended questions include a repository of additional questions that organisations can add to the core survey to tailor it to investigating gender equality in their unique context. The survey can be hosted on any LimeSurvey platform; the GEAM questionnaire, reporting template and documentation are open access.

The design of survey questions is as much a science as an art (Schaeffer & Dykema, 2011; Schaeffer & Presser, 2003). With this principle in mind, the research team deliberated over the design of questions, the provision of response options, the addition of "prefer not to say" options and open-text comment boxes, the inclusion of supplementary text within the survey and supporting guidance to accompany the GEAM. There was not necessarily a "right" or "wrong" answer to any of these decisions, as survey customs differ across disciplinary and national contexts. What guided our approach was the requirement for the GEAM to meet the needs of the ACT project and provide the CoPs with a tool to advance their work.

As a survey, the GEAM is primarily geared towards generating quantitative data about individual perceptions and experiences of the working environment. The GEAM is therefore envisaged as one element in a wider basket of data collection methods and to be used in conjunction with data from interviews and focus groups, document/discourse analysis and administrative data. The combination of quantitative data sources with qualitative data will provide a more comprehensive picture of known forms of exclusion and gender inequality as well as emergent topics not currently covered by standardised instruments such as the GEAM. Although primarily a tool to collect quantifiable data about individuals within research organisations, positioning the GEAM among other data collection methods responds to critiques of quantitative data from some feminist scholars and invites further reflections on what Sally E. Merry (2016) describes as the "seductions of quantification". Those who have highlighted the limitations of quantitative survey data tend to identify two major critiques, one epistemological and one methodological. From an epistemological position, survey instruments are historically associated with positivism (i.e., the view that knowledge is value-free and objective). Reflecting the scholarship of Donna Haraway (1988) and Helen E. Longino (1990), knowledge produced by the GEAM is situated within social contexts where the survey was designed and is conducted. In addition, with an eye to scholarship from the field of science and technology studies, John Law (2009) has noted how surveys can partly construct the categories they claim to observe. In other words, designing a survey that asks particular questions about gender equality in higher education and research organisations shapes how we come to define the borders and establish the scope of what is understood as gender equality.

In terms of methodology, scholars have questioned the measurement of gender, sex and sexuality in surveys and the tendency of this data collection method to present a reductionist account of the phenomena under investigation. For example, Anna Lindqvist et al. (2020, p. 1), addressing the measurement of gender, have noted that although much quantitative research in the social sciences uses gender as a variable, studies often overlook what is actually meant by gender and how is it operationalised in different studies. Amanda Baumle (2018, p. 281) has further described how a queer critique of quantitative data about gender, sex and sexuality might understand "categories as absent of real meaning". This chapter's case study presents a detailed account of how the GEAM responds to these critiques.

This chapter therefore positions the design of the GEAM survey within a wider scholarship that responds to challenges that come with the bridging of quantitative data and identity characteristics. These debates within gender studies and the broader social sciences have led us to qualify the use of the GEAM survey in the following ways. As a survey instrument, it must echo the approach of Ann Oakley (1999) who describes the power of quantitative methods to transform personal experiences in ways that demonstrate evidence of collective oppression. For example, describing the use of United Kingdom census data on sexual orientation and trans/gender identity to improve the lives of lesbian, gay, bisexual, trans and queer people in the United Kingdom, Kevin Guyan (2021, p. 8) has noted, "the census is a reflection for how some groups see themselves, present to others and transform identity characteristics into constituencies that form the basis for action to address inequality and injustice". The GEAM can produce knowledge about common experiences and perceptions among staff, with findings consequently used to inform the prioritisation of interventions. The GEAM intended to provide CoP members, as well as other higher education and research organisations, with a means of getting to "know their institution". However, the use of a survey to capture this information meant that the data collected about perceptions and experiences was frozen in time and space. Furthermore, although surveys shine a light on particular areas of an institution previously overlooked, coverage is not exhaustive and cannot necessarily cover the diversity of experiences related to gender inequality.

Knowing the institution – What is the purpose of the GEAM survey?

As well as collecting data on the identity characteristics of respondents, the GEAM also provides research organisations with a method to assess perceptions, beliefs and experiences across a range of topics that directly and indirectly relate to the theme of gender equality. This includes questions on parental leave, work-life balance, recruitment, promotion and working conditions during the COVID-19 pandemic. Unlike other gender equality reporting templates and indicators used at a European level, such as those developed by the Plotina project (Schwarz-Woelzl, 2015), modules target all employees and capture information about their perceptions and experiences of the working environment instead of institution-level data. The GEAM consists of five modules completed by individuals:

- Socio-demographics
- Working conditions
- Beliefs, attitudes and bias
- Organisational culture
- Behaviour and interpersonal relationships

Additional modules, such as institutional-level indicators and questions on COVID-19, were reviewed during the development and are available as extensions to the GEAM main questionnaire. The GEAM survey was primarily designed to facilitate and strengthen the work of the CoP member organisations, as part of the wider ACT project. For this reason, the GEAM needed to provide an online, adaptable questionnaire framework that produces comparable data and facilitates knowledge sharing, benchmarking and dialogue across national and organisational contexts. The research team wished to design something standardised – so that the same questions could be asked across different national/regional contexts – but also adaptable so that institutions could utilise the GEAM to advance gender equality in ways that work best in their context.

An innovative aspect of the GEAM is its technical setup and implementation. While many survey instruments are only available as Word documents or indirectly in scientific articles, the GEAM is available as a shareable LimeSurvey archive file. The modular approach makes it possible for users to import and setup a complete GEAM survey in a matter of minutes. Additional modular questions can be imported or exported as they become available on the dedicated portal database. In addition, the research team has produced a reporting template that generates descriptive statistics of a GEAM survey as a Word file. Using the popular software collaboration platform Github, GEAM users can contribute to all aspects of the project including its documentation, reporting templates and new questionnaire adaptations.[3] Although not originally part of the development plan, the research team has also developed a suite of manuals to accompany the GEAM, including: (i) an in-depth explanation of its development, documenting its reliability and validity; (ii) a practical, hands-on set of instructions that walks users through setting up the survey on LimeSurvey and handling the data generated responsibly; and (iii) an analysis handbook that provides users with an introduction to the statistical approaches that can be applied to the GEAM data, with examples of how to interpret these results. For more advanced users, the analysis handbook also includes the code required to run the prescribed analysis in the statistical software package R. Recognising that the majority of GEAM users are likely to be relatively unfamiliar with quantitative analysis, the research team also developed a reporting script that automatically generates an overview of the survey data as well as a breakdown of data by respondents' gender. Together, this suite of accompanying tools is intended to help those involved in gender equality research move from the initial step of identifying the issues to interpreting them in context and deciding which actions to pursue.

Although designed to facilitate and strengthen the work of the CoPs, the GEAM was also understood as an addition to a wider landscape of activity related to the collection and analysis of gender statistics. The United Nations (2006, p. 1) describes gender statistics as "statistics that adequately reflect differences and inequalities in the situation of women and men in all areas of life". Although described as "gender statistics", these activities entail the collection, analysis and presentation of data that is sex-disaggregated and gender-sensitive. Gender statistics are inclusive of statistics about topics that relate to gendered issues (such as maternity leave and childcare), reflect gender diversity (such as the proportion of women in senior leadership positions) and acknowledge social and cultural biases in the data (such as the use of collection methods that favour certain groups of people, for example, online staff surveys are only accessible to those that have access to the internet while at work). As noted by Engender (2020, p. 6), a Scottish feminist policy and advocacy organisation, the production of "gender statistics requires statisticians and analysts to move beyond simply counting women, and to fundamentally reconsider some of their assumptions about the world".

Following the 1995 Beijing Platform for Action, several programmes were introduced to expand and improve the collection, analysis and presentation of gender-sensitive data, with tools developed by institutions including the Organisation for Economic Co-operation and Development, UN Statistics Division and the World Bank (Engender, 2020, p. 5). Within the European Union, research projects funded under Horizon Europe will be required to have a Gender Equality Plan (GEP) that is public, supported by training and capacity building activities, has dedicated resources and engages in data collection and monitoring activities (European Commission, 2021). In specific national contexts, such as in Spain, there is a requirement for GEPs to discuss the collection, analysis and evaluation of initiatives by gender but no guidance or standardised suite of tools to support these requirements. In many cases, GEP implementation by different institutions has been conducted in isolation from others and with processes that differ widely in terms of their scope and effectiveness. In many cases, this implementation occurs without a proper assessment of gender equality needs and priorities, or the necessary monitoring and evaluation mechanisms (European Institute for Gender Equality, 2016). Results usually do not transcend the immediate project context while quality assessment of the reliability and validity of the generated data has not been conducted. It is at this juncture where the GEAM is intended to play a key role and provide a standardised assessment tool that advances mutual learning but is sensitive to local, national, regional and organisational contexts.

Designing the GEAM survey

Having discussed what the GEAM intends to achieve and the landscape of gender statistics within which the survey is situated, this section examines the origins and testing of the GEAM.

Origins

The origins of the GEAM are found in the Athena Survey of Science, Engineering and Technology (ASSET, see Aldercotte et al., 2017), a survey conducted in the United Kingdom at multiple time points (2003/04, 2006, 2010 and 2016) into the association between gender and experiences, expectations and perceptions of the workplace among academics in Science, Technology, Engineering, Mathematics and Medicine (STEMM). As a survey designed to assess the perceptions and experiences of researchers working within UK institutions, ASSET reflects the legislative context of equality, diversity and inclusion practice in the United Kingdom (most notably, for surveys conducted after 2010, the Equality Act and its articulation of nine protected characteristics (Equality and Human Rights Commission, 2021).

The ASSET 2016 survey contained 89 questions in total, including both categorical and continuous items (i.e., Likert scales). Questions asked in the ASSET 2016 survey were used to inform an initial framework for the GEAM. As part of ACT's second project meeting, in Berlin in 2018, the initial review was reviewed by Consortium members to identify any sensitive items and potential thematic gaps. This review also allocated items to one of six themes and 31 modules (see Table 3.1). This framework was further strengthened by a literature review (Aldercotte et al., 2021), which examined measurement scales used in previous studies on gender equality in different thematic contexts, and feedback from the wider ACT consortium.

Testing

After completion of an initial draft of the GEAM survey (i.e., version 1.0), items went through three rounds of assessment to ascertain their applicability across organisational contexts and countries:

1 **Focus groups:** The first draft of the GEAM survey was discussed in six focus groups carried out by seven partners in France, Germany, Sweden, Slovenia, Argentina, Spain and the United Kingdom to obtain feedback on the applicability of the GEAM across the unique institutional contexts of European countries as well as exploring its global transferability with a focus group in Latin America. Participants shared several observations including the need to not only capture empirical information about

Table 3.1 Framework and coverage of version 1.0 of the GEAM survey.

Socio-demographic variables	• Identity characteristics • Care responsibilities
Working conditions	• Job and career • Working arrangements and intensity • (Mental) health and safety • Job satisfaction
Beliefs, attitudes and stereotypes	• Perceived factors in career development • Perceived factors in recruitment and promotion • Sexism • Male/female identity and gender roles • Diversity • Leadership • Gender and status • Unconscious bias
Organisational culture and climate	• Perceptions of gender equality • Organisational culture and masculinity • Organisational climate • Working culture and career development • Barriers to training and career development • Parental leave experiences and culture • Campus climate • Group and team climate
Behaviour and interpersonal experiences	• Sexual harassment • Stalking and bullying behaviour • Bystander behaviour • Microaggressions • Interpersonal sexism

working conditions but also the working climate and the importance of ensuring an intersectional approach to data collection. In response to these points, questions on the perception of discrimination as well as the Masculinity Contest Culture were added. An extensive set of socio-demographic variables covering seven dimensions of social discrimination were also included in the refined, consolidated draft of the GEAM. Participants also identified what survey items to include as GEAM core survey questions and what to include as extended survey questions.

2 **Piloting with CoP members:** The consolidated draft of the GEAM was then implemented in the ACT LimeSurvey platform and piloted. Pilot participants were recruited principally through CoP facilitators with the aim to have one person per CoP member organisation responding. The pilot was launched in August 2019 and remained accessible online until the second week of September. A total of 68 responses were received. The resulting suggestions were integrated to produce version 1.0 of the GEAM core survey, which was then translated into Spanish, Polish, French and German for the final round of piloting.

3 **Piloting with non-CoP members:** After translating version 1.0 of the GEAM core survey, it was piloted with a group of CoP member institutions. The institutions distributed the survey to staff through their existing staff networks and had varying levels of response (ranging from around 60 participants to over 1,000 at the individual institutions). However, across all institutions there was a high degree of missing data, with large proportions of respondents dropping out of the survey after completing only a few modules, indicating issues with survey length. Taking into account the feedback on the length of the GEAM core survey and issues surrounding specific questions about respondents' gender, job and career and experiences of parental leave, version 2.0 of the GEAM core survey (see Table 3.2) was developed, updated and

Table 3.2 Version 2.0 of the GEAM core and extended GEAM survey (denoted by *)

Socio-demographics	• Age and marital status • Nationality and ethnicity • Religion* • Sex and gender • Disability • Education and income
Working conditions	• Job and career • Contract • Recruitment and promotion • Training • Caring responsibilities • Work-life balance/ work-family conflict scale • Job satisfaction • Work intensity • COVID-19 • Burnout and work engagement* • Lab safety*
Belief and bias	• Beliefs about unconscious bias* • Sexism* • Female/male identity and norms* • Leadership* • Diversity*
Organisational culture and climate	• Gender equality • Perceptions of work environment • Recruitment • Promotion* • perceived factors in career development • Masculinity contest culture • Team climate*
Behaviour	• Microaggressions • Bullying and harassment • Contra power*

finalised on the ACT LimeSurvey platform (for all language versions of the survey).

Following multiple rounds of testing, the GEAM survey includes five thematic blocks, containing a total of 43 modules and 54 core items and 112 extended items. The thematic blocks are ordered according to the following logic:

1 **Socio-demographic variables**: Includes information regarding respondents' age, gender, disability, sexual orientation, trans status/history, ethnicity and nationality as well as their socio-economic background.
2 **Working conditions:** Is a broad theme that contains items related to respondents' current contract, how they were recruited to their current role, their access to training opportunities and ratings of work-life balance, job satisfaction and work intensity.
3 **Belief and bias:** Focuses on social psychological constructs, i.e., it targets respondents' individual beliefs, attitudes and stereotypes regarding unconscious bias, sexism, masculine/feminine norms, diversity and leadership.
4 **Organisational culture and climate:** Extrapolates individual gender-related beliefs and attitudes towards the organisational or workgroup context. Its focus is largely on perceptions (targeted as "climate" measures) regarding gender equality but also deeper cultural aspects captured with the Masculinity Contest Culture scale. Climate refers primarily to how people perceive certain aspects of their work environment (e.g., teamwork) whereas culture has a normative dimension that captures employees underlying fundamental beliefs and values, for example, in relation to individualism, gender or respect for authority (Schneider et al., 2013).
5 **Behaviour:** Covers incidents of microaggressions, bullying and harassment and contra power harassment.

Case study: Asking about gender, sex, sexual orientation and trans status/history

One sticking point that emerged during the design, testing and roll-out of the GEAM survey was the measurement of the socio-demographic characteristics of gender, sex, sexual orientation and trans status/history. In particular, whether the GEAM should ask a single question presented as being about gender or two questions, one on gender and one on sex. The survey does not define the concepts of gender and sex (and the gender question does not explicitly mention "gender" in its question wording). It is therefore likely that many respondents will understand the two questions in the same way

and understand them as measuring the same concept.[4] The GEAM therefore includes the following three core questions:

Are you:

- A man
- Non-binary
- A woman
- Prefer not to say
- Other

Which best describes your sexual orientation?

- Bisexual
- Gay/lesbian
- Heterosexual/straight
- Prefer not to say
- Other

Are you trans or do you have a trans history?

- No
- Yes
- Prefer not to say

And the non-core question:

What is your sex?

- Female
- Male
- Prefer not to say
- Other

Except for "What is your sex?", the questions are core items and are therefore included as default in all versions of the GEAM. In addition, the
GEAM survey captures data about identity characteristics (including social
class, age and disability) to examine how perceptions and experiences of
gender equality intersect with other variables of social discrimination. The
research team were guided by the view that it is myopic to analyse gender equality by only collecting data about gender, and that perceptions and
experiences of gender are always shaped by how gender intersects with race
and other identity characteristics (Crenshaw, 1989).

The GEAM's adoption of a multi-dimensional approach to the collection of data about gender, sex and sexuality expands upon the approach of previous and ongoing EU-funded projects that measure gender equality, such as Plotina (Schwarz-Woelzl, 2015). Outputs from these projects collect data about "gender" but do not critically examine the interplay between gender, sex and sexuality. In addition to a multi-dimensional approach to gender, sex and sexuality data, the GEAM also collects information on respondent's age, marital status, ethnicity (majority or minority ethnic group), country of birth, citizenship, disability, education and parental education. The collection of data about several components of an individual's identity facilitates an intersectional approach to analysis, although the GEAM does not ask about race, religion or a detailed question on ethnicity.[5] For example, the disaggregation of GEAM results for respondents who identify as disabled, lesbian women or older, heterosexual men. Analysis at this granular level will depend on the size of samples but serves as an important reminder that perceptions and experiences of gender equality are intertwined with other facets of identity.

The GEAM, therefore, differed in its effort to respond to scholarship on gender, sex, sexual orientation and trans status/history data that highlights a lack of detail as to what we mean when we discuss gender and sex, the interchangeability of these concepts in the social sciences, and interactions between gender and sex and the related concepts of sexual orientation and trans status/history. For example, Amanda Bittner and Elizabeth Goodyear-Grant (2017, p. 1020) have described how "sex and gender tend to be used interchangeably among political behavior scholars (and others), both conceptually and operationally". As a result of the conflation of sex and gender, "gender also tends to be measured as a dichotomy, which is a normatively undesirable practice, and also one that inhibits precision in the measurement of gender" (Bittner & Goodyear-Grant, 2017, p. 1020).

In addition, scholars have noted that when surveys ask about sex, the concept is often naturalised in ways that assume something fixed and a binary separation between female and male respondents (Butler, 1990; Fausto-Sterling, 1993; Hawkesworth, 2013). Survey questions about sex often draw on biological criteria, which include asking about one's sex assigned at birth, and overlook how belief systems (in different cultures and time periods) have informed the meanings that societies attribute to different biological phenomena (Westbrook & Saperstein, 2015).

The GEAM survey aimed to avoid these mistakes and not to presuppose fixed or binary definitions of gender, sex or sexuality concepts. Furthermore, by demarcating core and extended survey items, the GEAM foregrounds asking a question about gender (with response options "man", "woman", "non-binary" and "other") above asking a question about sex (with response options "male", "female" and "other"). Lindqvist et al. (2020, p. 10) have argued that "researchers in the social sciences are rarely interested in the physiological/bodily aspects (i.e., genitalia, chromosomes, bodily attributes) or legal gender, but are more often interested in how individuals identify or express themselves from a social perspective". As a survey instrument designed to monitor gender inequality among those working in higher education and research contexts, the primary interest of the GEAM is phenomena associated with gender rather than sex, though the two concepts are interrelated.

The design of the GEAM's questions on gender and sex also had to acknowledge that some countries, such as Germany and the Netherlands, provide legal recognition to categories that transcend the binaries of man/male and woman/female. For example, in Germany, the constitution was amended to note the right for individuals to register their legal sex as "diverse" (Schotel & Mügge, 2021, p. 1). The provision of non-binary response options for questions on gender, sex and sexual orientation was unproblematic, particularly as research has highlighted that the inclusion of non-binary questions in population surveys does not invite adverse reactions from general respondents (Medeiros et al., 2020, p. 128).

The research team felt it was important to include a question that differentiated trans (people whose gender identity and/or expression is different from their sex as assigned at birth) and cis individuals (people whose gender identity and/or expression matches their sex as assigned at birth). The inclusion of a trans status/history question in the GEAM core survey also acknowledged that the perceptions and experiences of trans people have been historically overlooked (and in many instances, erased) from general studies as well as those specifically examining the topic of gender equality (Westbrook & Saperstein, 2015, p. 548). The GEAM survey uses an innovative approach to collect data on trans status/history that asks a single, direct question. Departing from the question format adopted by projects such as Gendered Innovations (European Commission, 2020, p. 193), the GEAM uses what is often described as a one-step approach to differentiate cis and trans respondents or those who have a trans history. In this example, the term "trans history" is understood to refer to people who no longer identify as trans but were assigned a sex a birth that differs from how they currently identify (in other words, they have transitioned). The GEAM asks one question to identify this population, "Are you trans or do you have a trans history?" This approach differs from a two-step method that asks one question about an individual's sex at birth and one question about their current gender identity. When viewed together, this data is used to discern who might identify as trans. Within the context of the United Kingdom, the use of a two-step approach is problematic due to the Gender Recognition Act. There are also concerns about the acceptability or relevance of asking people about their sex at birth in a survey designed to gauge their present perceptions and experiences of gender equality, this observation might be particularly true of trans people and those who are intersex or have differences in sex development. Furthermore, the European Commission (2020, p. 193) has noted difficulties in the use of a two-step question when translated into languages that use a single term to refer to both sex and gender identity (the Commission notes Danish, Norwegian and Sweden as examples).

Translation and comprehension

Throughout the design of the GEAM survey, the research team were mindful of construct equivalence – the view that measurable concepts exist that are understood by all groups answering a survey. For example, it is difficult if not impossible to make sense of responses to a question about "bullying and harassment" if respondents understand this concept in different ways. As Heather Ridolfo et al. (2012, p. 117) explain, "data lacks comparability if a particular construct does not exist, or is fundamentally different, in one or more of the represented cultural groups".

As the GEAM was to be translated into multiple languages and used in a diversity of institutions and countries, the need for equivalent constructs was vital to allow for the possibility of analysis across different contexts. Translations were revised or implemented from scratch by Consortium partners or directly by external individuals interested in carrying out a GEAM survey in their organisation. In either case, translators experienced in gender equality made sure that the nuances and meaning of the English concepts used were adequately captured in the destination language. Since many translations were voluntary contributions, reverse translation from the destination language to the source (English) language was not carried out.

The challenge of construct equivalence did not only relate to the comprehension of topics associated with gender equality but also the socio-demographic terms used to describe the gender, sex and sexuality characteristics of respondents. In the United Kingdom, for example, the Office for National Statistics (ONS) has described its difficulty in translating questions about gender from English into other languages, particularly when questions are intended to differentiate concepts of gender and sex (UN Economic and Social Council, 2019, p. 5). During the design of the question on gender identity for the 2021 English and Welsh census, the ONS reported that it was not possible to adequately translate this concept into German, Dutch, Romanian or Greek (as it had to differ from another census question that asked about sex) (ibid.). As Leslie W. Suen et al. (2020, p. 2313) observe, "depending on the translation and cultural variations, participants could be self-categorizing themselves differently or using the write-in option different from originally intended" by the designers of the survey.

A recurring theme in feedback received on the GEAM were flashpoints where data collected was not applicable to the context under investigation. These flashpoints ranged from what appeared at first to be relatively simple concepts, such as the labels that respondents ascribed to their current contract (e.g., the term academic was widely recognised but the labels associated with administrative and support staff varied considerably across pilot contexts), to more nuanced issues such as the availability of various resources and support surrounding parental leave and childcare that vary considerably across legal contexts and may not be at the discretion of the institution to offer staff.

Reception and acceptability

The research team designed the GEAM with an eye towards its reception and the acceptability of questions asked among respondents. As participation was voluntary, the design of the GEAM erred in favour of asking more questions (and possibly more probing questions) rather than less. This decision was based not only on the research team's personal experiences of conducting questions that ask about gender, sex and sexuality but also a wider literature on the reception of questions, particularly within the context of the United States and LGBT communities. The Williams Institute (Badgett, 2009), a think tank based at the University of California in Los Angeles, has challenged the belief that asking questions about the perceptions and experiences of LGBT people is a sensitive topic or likely to distress respondents in mainstream population studies that might risk them ending their participation in the survey. Although it is vital that survey questions about gender, sex and sexuality are presented in a considered and respectful manner, for most respondents these topics are not understood as taboo and – when asked – most people are willing to provide an answer. Scholarship has also noted that, depending on the design of the study, individuals from sexual minority groups are not necessarily less forthcoming or harder to reach than those from sexual majority groups. Nancy Bates et al. (Bates et al., 2019, p. 718), for example, has noted that "despite this implicit assumption, there is little empirical evidence on the topic".

Discussion

The GEAM not only functions as a means to "know the institution" but also has a normative function in determining what is brought into view and what is precluded from examinations of gender equality in higher education and research organisations. Knowingly or unknowingly, surveys on gender equality can come to describe the phenomena they claim to observe – not only through the data collected but through the design and selection of questions. The effects of research methods on the problems they are intended to investigate have been explored by several scholars. Law (2009, p. 239), for example, explains, "there are two great views of method in science and social science. On the one hand, it is usual to say that methods are techniques for describing reality. Alternatively, it is possible to say that they are practices that do not simply describe realities but also tend to enact these into being". If the GEAM is understood as a means to bring certain topics or issues into being, it can also function as a tool to raise awareness about gender equality within institutions (for example, among senior leaders).

As a means to generate standardised quantitative data, the GEAM can also be contextualised with an overall trend in research organisations towards governance through benchmarking and quantification. As Merry (2016) argues, indicators and quantitative data have a knowledge effect (i.e., facts about pay gap) as well as a governance effect (i.e., they provide a basis for decision making and accountability). The production of harmonised statistics and quantitative indicators prepares the ground for competitive benchmarking exercises, which suggest the replacement of politicised decision making with supposedly "fair" and "neutral" standards of statistical quantification (Bruno, 2009). In light of this fiction, it is vital to remain aware of the constructed nature of survey instruments and indicators, which – once designed – tend to acquire a life of their own that hides the ambiguities and subjective decisions that informed its construction.

There are therefore clear limitations as to what can be brought into being through the methods selected, as the normative functions of the GEAM need to reflect existing gender equality discourses. For example, the GEAM would not be understood or well-received by respondents if it asked a question about the ability of women to lead research projects as this stereotype is out-of-step with current views in most contexts where the survey was conducted. It is therefore more helpful to consider the GEAM as an output that can shine a light on some issues and not others. There is much to learn when we view the GEAM as doing more than providing a mirror image of gender equality in the higher education sector but also informing what is included and excluded in this reflection. We hope that these conceptual reflections on the design of the GEAM bring some of these discussions into view in ways that encourage critical reflection among others engaged in the design and roll-out of gender equality surveys.

As the ACT project concluded, the roll-out of the GEAM survey highlighted areas where ideas about gender equality, education and research, and working practices are conceptualised differently – both in terms of language and how ideas are understood across borders. In particular, the collection of socio-demographic data about individual respondents required the design of standardised diversity monitoring questions, response options and supplementary guidance that were understood across different national contexts. Although the research team engaged in critical reflections as to the potentials and limitations of the GEAM survey throughout its design and multiple iterations, it was also apparent that the tool could not deliver everything expected of it. Returning to the socio-demographic measurement of gender, sex and sexuality discussed in this chapter's case study, critics of the GEAM might highlight the limitations of the approach adopted. For example, although focused on gender equality, the GEAM survey does not provide insights into how gender expression might affect people's perceptions and experiences. This gap is not unique to the GEAM; Lindqvist et al. (2020, p. 8) note that gender expression is seldom "accounted for in the social sciences" and, citing the work of Devon Magliozzi et al. (2016), explain that it is rare to see surveys that ask "participants about how feminine and masculine they see themselves, and how feminine and masculine they believe others see them". Furthermore, although the GEAM attempts to provide a more rounded account of gender, sex and sexuality through the provision of multiple questions, the provision of categorical options for all four questions means that potential insights are perhaps missed.[6] Although the decision was made to ask questions that challenge binary assumptions about gender and sex, the research team did not consider the inclusion of a question that captured information about gender as a continuum as this did not align with approaches to gender equality used in other parts of the ACT project and risked confusing GEAM survey respondents.

There is also the question as to what happens with data collected via the GEAM survey. Moving from data to action – early feedback suggests challenges around who undertakes the analysis of data collected and its use to inform action within an institution. The team involved in the GEAM have therefore had to ensure we provide the right level of support to institutions so that we are not too prescriptive but equally able to empower institutions to use the GEAM survey for action. For example, in its support of the CoPs, the GEAM can also collect data that informs an evidence base for action. This evidence base might include the use of quantitative data to convince senior leaders of the existence of problems and the need to take action or insights into what works (and what does not work) to improve gender equality in different contexts.

While the GEAM has some limitations (for example, the categorical approach to identifying respondents' gender), it nonetheless provides CoPs and others engaged in addressing gender inequality with a starting point and helps move the dial in the direction of progress. As Baumle (2018, pp. 281–282) describes, strength comes from disclosing the weaknesses and limitations of the data collected while also "emphasizing the importance of generating new knowledge". In regard to the remit of the ACT project, the GEAM has provided institutions with a tool to collect data that demonstrates the existence of gender inequality where previously no evidence base existed.

Conclusion

As a research instrument translated into multiple languages and rolled out across different national higher education and research contexts, the GEAM survey presents an insight into some of the challenges associated with the transnational measurement of gender, sex and sexuality. A key strength of the GEAM survey is that the items have been sourced from existing surveys that have been used previously to investigate gender differences in staff experiences and perceptions. The items have been adapted and tested for use across a range of contexts, lending a greater deal of assurance of their reliability and validity than previous gender equality surveys. The paper concludes with a discussion of the tensions between the design of a universal and culturally-specific measurement tool, as well as the overlap and interplay between concepts of gender, sex and sexuality and early indications of how this multi-dimensional account presents a more detailed picture of respondents' experiences of gender equality across higher education and research contexts.

Notes

1. Communities of PrACTice for Accelerating Gender Equality and Institutional Change in Research and Innovation across Europe" Horizon 2020 project, grant number 788204 is referred to throughout this book as "The ACT project". See also https://www.act-on-gender.eu.
2. For details and documentation regarding the GEAM, please visit https://geam.act-on-gender.eu.
3. See https://github.com/actongender.
4. It is estimated that around 99% of the United Kingdom population perceive or experience no difference between their gender and sex (Government Equalities Office, 2018). Furthermore, in regard to how sex is conceptualised in United Kingdom censuses, the Office for National Statistics has described how there exists five concepts of sex that a data collection exercise could ask about: sex as registered at birth; sex as recorded on birth certificate; sex as recorded on legal/official documents; sex as living/presenting; and sex as self-identified (Rosiecka, 2021).
5. It is uncommon (and in many contexts illegal) to capture data about a person's race or ethnicity in countries where the GEAM survey was implemented.
6. In other words, the GEAM does not use continuum scales to measure gender identity. The provision of scales to measure the gender of respondents might take the form of a bipolar scale (man/masculine ←→ woman/feminine) or two unipolar scales (man/masculine ←→ not man/masculine, woman/feminine ←→ not woman/feminine) (Gidengil & Stolle, 2021, p. 2).

References

Aldercotte, Amanda, Caprile, Maria, Guyan, Kevin, Malibha-Pinchbeck, Memory, Müller, Jörg, Palmén, Rachel, & Startin, Carla (2021). *ACT – Gender Equality Audit and Monitoring (GEAM) – Version 2*. Zenodo. https://doi.org/10.5281/zenodo.5348197

Aldercotte, Amanda, Guyan, Kevin, Lawson, Jamie, Neave, Seave, & Altorjai, Szilvia (2017). *ASSET 2016: Experiences of gender equality in STEM academia and their intersections with ethnicity, sexual orientation, disability and age*. Equality Challenge Unit. http://www.ecu.ac.uk/wp-content/uploads/2017/04/ECU_ASSET-report_April-2017.pdf

Badgett, M. V. Lee (2009). *Best practices for asking questions about sexual orientation on surveys*. The Williams Institute. https://escholarship.org/uc/item/706057d5

Bates, Nancy, García Trejo, Yazim A., & Vines, Monica (2019). Are Sexual Minorities Hard-to-Survey? Insights from the 2020 Census Barriers, Attitudes, and

Motivators Study (CBAMS) Survey. *Journal of Official Statistics*, *35*(4), 709–729. https://doi.org/10.2478/jos-2019-0030

Baumle, Amanda K. (2018). The Demography of Sexuality: Queering Demographic Methods. In D. Compton, T. Meadow, & K. Schilt (Eds.), *Other, please specify: Queer methods in Sociology* (pp. 277–290). University of California Press.

Bittner, Amanda, & Goodyear-Grant, Elizabeth (2017). Sex isn't Gender: Reforming Concepts and Measurements in the Study of Public Opinion. *Political Behavior*, *39*(4), 1019–1041. https://doi.org/10.1007/s11109-017-9391-y

Bruno, Isabelle (2009). The "Indefinite Discipline" of Competitiveness Benchmarking as a Neoliberal Technology of Government. *Minerva*, *47*(3), 261–280. https://doi.org/10.1007/s11024-009-9128-0

Butler, Judith (1990). *Gender trouble: Feminism and the subversion of identity.* Routledge.

Crenshaw, Kimberle (1989). Demarginalizing the Intersection of Race and Sex: A Black Feminist Critique of Antidiscrimination Doctrine, Feminist Theory and Antiracist Politics. *University of Chicago Legal Forum*, *1989*(1), 139–167.

Engender. (2020). *Sex/gender: Gathering and using data to advance women's equality and rights in Scotland* [Engender Submission to the Office of the Chief Statistician]. Engender. https://www.engender.org.uk/content/publications/EgenderSubmission-GatheringandusingdatatoadvancewomensequalityandrightsinScotland-Feb2020-1.pdf

Equality and Human Rights Commission. (2021). *Protected characteristics.* https://www.equalityhumanrights.com/en/equality-act/protected-characteristics

European Commission. (2011). *Structural change in research institutions: Enhancing excellence, gender equality and efficiency in research and innovation.* Publications Office of the European Union. https://doi.org/10.2777/94044

European Commission. (2020). *Gendered innovations 2: How inclusive analysis contributes to research and innovation: policy review.* Publications Office of the European Union. https://data.europa.eu/doi/10.2777/316197

European Commission. (2021). *Horizon Europe guidance on gender equality plans.* Publications Office of the European Union. https://data.europa.eu/doi/10.2777/876509

European Institute for Gender Equality. (2016). *Gender equality in academia and research GEAR tool.* Publications Office of the European Union.

Fausto-Sterling, Anne (1993). The Five Sexes. *The Sciences*, *33*(2), 20–24. https://doi.org/10.1002/j.2326-1951.1993.tb03081.x

Gidengil, Elisabeth, & Stolle, Dietlind (2021). Comparing Self-Categorisation Approaches to Measuring Gender Identity. *European Journal of Politics and Gender*, *4*(1), 31–50. https://doi.org/10.1332/251510820X15918093444206

Government Equalities Office. (2018). *National LGBT survey. Research report (GEO–RR001).* Government Equalities Office. https://assets.publishing.service.gov.uk/government/uploads/system/uploads/attachment_data/file/721704/LGBT-survey-research-report.pdf

Guyan, Kevin (2021). Constructing a Queer Population? Asking about Sexual Orientation in Scotland's 2022 Census. *Journal of Gender Studies*, *0*(0), 1–11. https://doi.org/10.1080/09589236.2020.1866513

Haraway, Donna (1988). Situated Knowledges: The Science Question in Feminism and the Privilege of Partial Perspective. *Feminist Studies*, *14*(3), 575–599. https://doi.org/10.2307/3178066

Hawkesworth, Mary (2013). Sex, Gender, and Sexuality: From Naturalized Presumption to Analytical Categories. In G. Waylen, K. Celis, J. Kantola, & S. L. Weldon (Eds.), *The Oxford handbook of gender and politics*. Oxford University Press. https://doi.org/10.1093/oxfordhb/9780199751457.013.0001

Law, John (2009). Seeing Like a Survey. *Cultural Sociology, 3*(2), 239–256. https://doi.org/10.1177/1749975509105533

Lindqvist, Anna, Sendén, Maria G., & Renström, Emma A. (2020). What is Gender, Anyway: A Review of the Options for Operationalising Gender. *Psychology & Sexuality, 0*(0), 1–13. https://doi.org/10.1080/19419899.2020.1729844

Longino, Helen E. (1990). Science as social knowledge. In *Science as social knowledge*. Princeton University Press. https://www.degruyter.com/document/doi/10.1515/9780691209753/html

Magliozzi, Devon, Saperstein, Aliya, & Westbrook, Laurel (2016). Scaling Up: Representing Gender Diversity in Survey Research. *Socius, 2*, 2378023116664352. https://doi.org/10.1177/2378023116664352

Medeiros, Mike, Forest, Benjamin, & Öhberg, Patrik (2020). The Case for Non-Binary Gender Questions in Surveys. *PS: Political Science & Politics, 53*(1), 128–135. https://doi.org/10.1017/S1049096519001203

Merry, Sally E. (2016). *The seductions of quantification. Measuring human rights, gender violence, and sex trafficking.* The University of Chicago Press.

Oakley, Anne (1999). Paradigm Wars: Some Thoughts on a Personal and Public Trajectory. *International Journal of Social Research Methodology, 2*(3), 247–254. https://doi.org/10.1080/136455799295041

Ridolfo, Heather, Miller, Kristen, & Maitland, Aaron (2012). Measuring Sexual Identity Using Survey Questionnaires: How Valid Are Our Measures? *Sexuality Research and Social Policy, 9*(2), 113–124. https://doi.org/10.1007/s13178-011-0074-x

Rosiecka, Helena (2021). *Methodology for decision making on the 2021 census sex question concept and associated guidance.* UK Statistics Authority. https://uksa.statisticsauthority.gov.uk/publication/methodology-for-decision-making-on-the-2021-census-sex-question-concept-and-associated-guidance/

Schaeffer, Nora C., & Dykema, Jennifer (2011). Questions for Surveys: Current Trends and Future Directions. *Public Opinion Quarterly, 75*(5), 909–961. https://doi.org/10.1093/poq/nfr048

Schaeffer, Nora C., & Presser, Stanley (2003). The Science of Asking Questions. *Annual Review of Sociology, 29*(1), 65–88. https://doi.org/10.1146/annurev.soc.29.110702.110112

Schneider, Benmain, Ehrhart, Mark G., & Macey, William H. (2013). Organizational Climate and Culture. *Annual Review of Psychology, 64*(1), 361–388. https://doi.org/10.1146/annurev-psych-113011-143809

Schotel, Anne Louise, & Mügge, Liza (2021). Towards Categorical Visibility? The Political Making of a Third Sex in Germany and the Netherlands. *JCMS: Journal of Common Market Studies, 59*(4), 981–1024. https://doi.org/10.1111/jcms.13170

Schwarz-Woelzl, Maria (2015). Plotina Monitoring Tool. *Plotina Project.* https://www.plotina.eu/monitoring-tool/

Suen, Leslie W., Lunn, Michel R., Katuzny, Katie, Finn, Sacha, Duncan, Laura, Sevelius, Jae, Flentje, Annesa, Capriotti, Matthew R., Lubensky, Micah E., Hunt, Carolyn, Weber, Shannon, Bibbins-Domingo, Kirsten, & Obedin-Maliver, Juno (2020). What Sexual and Gender Minority People Want Researchers to Know About Sexual Orientation and Gender Identity Questions: A Qualitative Study.

Archives of Sexual Behavior, *49*(7), 2301–2318. https://doi.org/10.1007/s10508-020-01810-y

UN Economic and Social Council. (2019). *In-depth review of measuring gender identity*. Conference of European Statisticians, Paris. https://unece.org/fileadmin/DAM/stats/documents/ece/ces/2019/ECE_CES_2019_19-G1910227E.pdf

United Nations (Ed.). (2006). *The world's women 2005 Progress in statistics*. United Nations.

Westbrook, Laurel, & Saperstein, Aliya (2015). New Categories Are Not Enough: Rethinking the Measurement of Sex and Gender in Social Surveys. *Gender & Society*, *29*(4), 534–560. https://doi.org/10.1177/0891243215584758

4 Co-creation methods for Communities of Practice

Towards institutional change

Aleksandra Thomson, Kathrin Rabsch,
Sarah Barnard, Tarek M. Hassan
and Andrew R. J. Dainty

Introduction

Communities of Practice (CoP) have been extensively used and investigated as an approach for organisational learning and knowledge transfer, however, applying CoPs as an instrument for *institutional change* from a facilitation perspective has been underexplored (Hogan, 2002). References on how to support CoPs which aim to foster institutional change remain rare. One notable exception found in the literature focuses on facilitating group decision making processes through the so-called causal mapping method to improve productivity (Pyrko et al., 2017), however, there is a lack of more specific ground level tools and techniques that could be utilised by practitioners involved in cultivating CoPs for institutional change. This is surprising since CoPs are complex, contextually sensitive social phenomena that must evolve and develop over time to fulfil their potential; thus, they require a facilitative framework to connect, work and "think together" (Pyrko, Dörfler, & Eden, 2017, p. 390). Collaboration and cooperation are crucial CoP success facets and if communities are intentionally organised for institutional change for equality and broader social justice, then the CoPs will need to be robustly concerted, collective, and transformative, which can be effectively fostered through *co-creation*.

   Co-creation activities are "practices where actors engage collaboratively [...] through interactions within a specific social context" (Frow et al., 2015, p. 26). The aim of co-creation is to collaborate, create together, cooperate, and share ideas, knowledge, practice, and build on the existing ideas to develop them further. An example of co-creation is recasting service, policy or product users as "co-producers" and inviting them to the designing, planning, and delivery processes, and in creating outputs collaboratively (Osborne & Strokosch, 2013). Facilitation is particularly conducive to co-creation activities, since it "is concerned with encouraging open dialogue among individuals with different perspectives so that diverse assumptions

DOI: 10.4324/9781003225546-4

and options may be explored" (Hogan, 2002, p. 10). Successful and fruitful co-creation also thrives on an equal contribution from the members of the CoPs and from incorporating the diversity of voices and perspectives. This multiplicity of collaboration partners is thus particularly desirable in gender equality projects or change programmes within higher education (HE) and research and innovation, as it reflects the spirit of equality, diversity, and inclusion, and its values of fairness and opportunity for all (Acker, 2000). However, it must be acknowledged that a potential paradox in facilitation can occur, namely, that the influence required to facilitate a community can change the group's outcomes. Nevertheless, strict impartiality for facilitation may not be possible as every human actor brings their own biases and agenda into the group (Griffith, Fuller, & Northcraft, 1998). This must be recognised, and both the advantages and disadvantages of such dynamics reconciled at the community level.

The benefits of using co-creation vary from harnessing the active involvement of participants in co-creating thus securing "buy-in", sharing resources and knowledge to enhancing innovation processes, providing network solutions, and contributing to the well-being of the service system (Frow et al., 2015). As the foundation of co-creation is participation and collaboration, the use of participatory methods for consensus building, sharing experiences and mentoring are helpful tools to support CoPs in operating, developing, and implementing gender equality plans (GEPs) and broader equality interventions in institutions.

This chapter is structured as follows: we first draw on the extant theoretical literature on CoPs (i.e., classical structural model, lifecycle phases, areas of activity, success factors) that inform our selection of co-creation methods. Second, building on the synthesis of these literatures, we present the participative methods curated for the ACT Co-Creation Toolkit (accessible online) and we encourage the reader to access the Toolkit for more in-depth technical information about the methods (Thomson & Rabsch, 2021). The Toolkit was developed as a deliverable for the ACT-on-Gender project to support CoP facilitators and members in institutional change efforts towards gender equality in HE and research organisations.

Throughout the chapter, we refer to the methods contained in the full version of the Toolkit; however, we specifically focus on four activities in more detail: *DAKI Retrospective*, *1-2-4-All*, *Plan of Change*, and *Future Workshop*. The reason for selecting these methods is that they arguably address the most pertinent needs of CoPs embarking on gender equality projects: they need to be inclusive, allow diversity of perspectives, and facilitate collaborative imagination of the future and the planning of change. The following aspects of the methods are covered: the aims, the preparation, and the structure of the method. The chapter concludes with the discussion on how the proposed co-creation methods contribute to the practice of cultivating CoPs that focus on institutional change.

Communities of (Political) Practice: The structural model

CoPs are defined as groups of people who share a concern for the same problem or topic, and who on that basis engage to learn together and from each other (Wenger, McDermott, & Snyder, 2002). Whilst this definition strongly frames our understanding of CoPs, we draw on the definition of CoP as a political endeavour (see Chapter 2) to sensitise us to the distinct needs of such communities, i.e., Communities of Political Practice (CoPPs) are groups of "institutionally affiliated people across different organisations or nations coalescing around a shared concern for *social equality* who engage in *transformative practice*, who learn from each other and co-create knowledge through regular interactions to act on *institutional change*" (p. 26, this volume). As such, when we refer to "CoP", we do so to discuss the extant CoP theory and its classical building blocks, and when we refer to "ACT CoP", we do so to emphasise the unique context of CoPs for *transformative aims* within institutions.

Wenger, McDermott, and Snyder (2002) defined three structural elements of CoP: *domain, community,* and *practice.* The *domain* is the topic a CoP is invested in and to which the members are committed, and the ACT CoPs' domain is broadly centred around gender equality in research organisations. The *community* creates the social fabric of learning where belonging, collective identity, and intellectual processes are fostered. ACT CoP comprises academics, researchers, and practitioners collectively invested in the shared domain of gender equality in the affiliated research or HE institutions. The *practice* is a set of frameworks, ideas, tools, and resources, which CoP develops, shares, and maintains (Wenger, McDermott, & Snyder, 2002); whilst for the ACT CoPs, such practice has a transformative aim to enable the communities to progress efficiently and promote and accelerate institutional change, which have been conceptualised as CoPPs (see Chapter 2).

As such, cumulatively, the needs of CoPs for institutional change are nuanced. Co-creation methods will need to emphasise the involvement of diverse voices, perspectives and contexts and create opportunities for equal and fair engagement. Moreover, foregrounding gender equality as a change agenda can be a perilous endeavour (van den Brink, 2020), requiring vision and tools for transformation. It is important to add that the suggested methods contained in the Toolkit can be used at any point in the community development as needed, and that many methods lend themselves to more than one CoP lifecycle phase, area of activity, or approach to building a CoP. Our aim is to sensitise CoP facilitators and CoP managers to the diverse aspects of community building and suggest certain methods that may be useful throughout the CoPs' lifecycle phases.

Co-creation methods for CoP lifecycle phases

McDermott (2000) proposed that CoPs have six lifecycle phases, and that specific design, facilitation, and support strategies exist to help reach the goals of the CoP during each phase and raise it into the next stage of development.

The *inquire* phase is characterised by the efforts of exploration and inquiry with the key stakeholders and determining the goals and the vision for the CoP. This phase is followed by the *design* phase which focuses on defining activities, processes, and different roles that support the goals of the community. These fundamental aspects allow the community to move forward. In the *prototype* phase, the community builds the necessary commitment, refines strategies, tests assumptions, and creating a workable success story takes place. Next, the *launch* phase begins as the community is planned and structured and can present itself to a broader audience and engage other stakeholders. Once established, the CoP begins its *grow* phase and develops by engaging more members, participating in events, and reaching the first established goals. The main activities are focused on learning collaboratively, sharing knowledge, engaging in group projects, and networking events, while creating an increasing cycle of participation and contribution. If the community is to survive, it is essential to strengthen it by assessing what has been achieved and planning the next steps. Therefore, in the *sustain* phase new goals are being set and new strategies developed that build on previous experiences to prevent the CoP from disbanding. Co-creation methods that acknowledge the lifecycle phases of CoPs need to be nuanced and designed to foster each stage of development.

The first two phases (*inquire*, *prototype*) are characterised by exploration, investigation, and defining of new activities, roles, and processes. They are more flexible and open to possibilities and opportunities, thus especially important, as they ultimately shape the community's audience, goals, vision and how it will choose to operate (Cambridge, Kaplan, & Suter, 2005). For the *inquire* phase, we suggest suitable co-creation methods including idea generation activities (e.g., *Brainstorms*; *1-2-4-All*; *SWOT/PESTEL analyses*), team building (e.g., *Five-Minute Favour*; *Four Quadrants*; *Stinky Fish*; *What I Need from You*); and for the *design* phase, vision building (e.g., *Future Workshop*; *How, Now, Wow*), and change planning (e.g., *Critical Uncertainties*; *DAKI Retrospective*; *Plan of Change*). CoPs that aim to foster institutional change will be particularly attentive to relevant co-creation change methods during these two cycles, since they will need to collaboratively envisage a desirable future, but also unpack how to reach the goals, and what tools, support, and resources they may require on the change journey. It is also crucial for them to be inclusive and involve diverse voices and perspectives in this vision building.

In the *prototype* and *launch* phases, the formal activities are piloted and then become fully operational. Collectively, decisions must be made about selecting the most appropriate technical features to support the goals of the CoPs, facilitating synchronous and asynchronous events and activities, ensuring the clarity of roles and support structures being in place, and setting up communication channels. Some key questions will need to be answered by the whole community as to the gatekeeping and joining the CoPs, the benefits of becoming a member, what kinds of community activities will generate the desired energy and engagement, and how to best support the emergence of community presence (Cambridge et al., 2005). To aid these phases, we suggest co-creation methods that promote peer support for the new and established members (e.g., *What I Need from You*; *Five-Minute Favour*; *Mentoring Circles*), but also to encourage reflecting how different stakeholders could benefit not only from the CoP membership directly, but also from its activities and the planned change indirectly (e.g., *Personas*).

The *grow* phase is based around engaging CoP members and continuing what has already been established, such as regular facilitation and communication. It is useful to create and share success stories within the community (e.g., by using techniques such as storyboards, photo documentation, or simple digital stories) to showcase best practices and create a sense of thriving. In this phase, it might be also useful to reflect on the activities and achievements so far to identify what has worked and what has not been effective or beneficial (e.g., *DAKI Retrospective*). To facilitate discussions about the community itself, the community culture, processes and practices, technology, and individual motivations for participating (Cambridge et al., 2005), co-creation methods that elicit deeper meaning might be useful to adopt (e.g., *Fish Bowl*; *Argument Mapping*; and *W3: What, So What, Now What*).

The final *sustain* phase is characterised by cultivating and assessing the learning, knowledge, and outputs which have been created by the communities and reflecting on how they inform new strategies, aims, activities, roles, and plans (Cambridge et al., 2005). Therefore, it is useful to draw on co-creation methods that help to identify what has been achieved and what potential challenges lay ahead for the community (e.g., *Critical Uncertainties*; *SWOT/PESTEL analyses*; *DAKI Retrospective*).

Co-creation methods for the main areas of CoP activity

The main areas of CoP activity proposed by McDermott (2000) are already strongly emerging in the *launch* lifecycle phase of a CoP. These are categorised as *building relationships*; *learning and developing practice*; *taking action as a community*; and *creating knowledge in the domain*. These activities do not map directly onto the lifecycle phases, but rather they cut across all phases of CoP development, e.g., in the *sustain* phase, CoPs may wish to focus on building new relationship to open up new membership opportunities to prolong the community. However, Cambridge et al. (2005) suggest that beyond the CoP lifecycles, each community is characterised by their unique goals, purpose and the members' characteristics and needs. Therefore, it is important that all social and technical design choices are primarily driven by purpose and the context of the CoPs. Communities that succeed and that last are characterised by focused and well-defined purposes that are linked to the strategic mission of the sponsoring organisation. The most effective way to define a CoP's purpose is to assess how this initiative will benefit the community's stakeholders and what specific needs are to be met by the community.

The first area of activity, *building relationships,* refers to the interaction with others and the development of a network with other people dedicated to the same topic, thus through these relationships, the commitment of the community will be strengthened. Therefore, we suggest facilitators choose participative methods that support continuous and deep interactions between the CoP members (e.g., *Five-Minute Favour*; *Four Quadrants*; *Heart, Hand, Mind*).

Second, as *learning and developing practice* is a collective process within a community, it builds on the extant knowledge within the individual members, and it requires co-creation activities that stimulate sharing, building common meaning and understanding, and creativity to adapt and innovate to take the practice forward. To facilitate these key processes, we recommend activities that are inclusive, encourage participation, and idea maturation (e.g., *1-2-4-All*; *Mature Your Ideas*).

The third main aim of a CoP is to make things happen, to drive change, and to *take action as a community* through tasks and projects. This is particularly relevant in CoPs focusing on institutional change. Working with others and collaboration towards these goals can be supported by activities that focus on preparing an effective response to potential future challenges and developing a strategy with an implementation plan (e.g., *Critical Uncertainties*; *Plan of Change*; *SWOT/PESTEL analyses*).

Lastly, to achieve change and progress, CoPs will need to transcend the status quo of the area of interest by cross-pollinating ideas and spreading expertise to innovate and *create knowledge in the domain.* Co-creation activities should therefore support the propagation of ideas and knowledge (e.g., *Argument Mapping*; *Mentoring Circles*; *the World Café*).

Co-creation methods for the CoP success factors

There is rich scholarship available on CoP success and failure factors, and the authors of this chapter conducted a literature search in peer-reviewed, full-text journals and conference papers that included phrases "communit* of practice", "success factor", and "failure factor". The nine most recurring success factors were identified (see Table 4.1) to guide the development of the ACT Co-creation Toolkit and the choice of the included methods.

Mutual learning, sharing, and producing knowledge are the key features of successful communities thus these processes are crucial for the CoP development and accomplishing its goals (Hong, 2017). Whereas learning focuses mostly on CoPs internal activities through regular meetings and learning from each other, knowledge production and access to knowledge is contingent on the value of inviting external experts to import knowledge and new perspectives. This can also be facilitated by connecting with other networks or working groups focused on a similar topic, which will allow the CoP to create new knowledge and practices based on the input provided from outside the CoP (Probst & Borzillo, 2008). This, of course, also applies

Table 4.1 Community of Practice success factors and the identified literature.

Success factor	Example	Identified literature
Community interaction	Community support People factors Engagement and participation Regular interaction and communication	Akhavan, Marzieh, & Mirjafari (2015) Cambridge et al. (2005) Fontainha & Gannon-Leary (2008) Jagasia, Baul, & Mallik (2015) McDermott (2000) Probst & Borzillo, (2008) Pyrko, Dörfler, & Eden (2017) Sanz Martos (2012)
Sharing best practice	Saving resources (time & budget) Personal learning	Hong (2017); Probst & Borzillo (2008); Retna & Ng (2011);
Supporting tools and resources	Provision of infrastructure, tools, technology, premises	Akhavan et al. (2015); Fontainha & Gannon-Leary (2008); Hong (2017);
Mutual culture, values, belonging	Shared culture and vision Mutual engagement Belonging Trust, common values Shared understanding, cultural awareness	Fontainha & Gannon-Leary (2008); Pyrko, Dörfler, & Eden (2017); Retna & Ng (2011); Sanz Martos (2012);
Knowledge production and access to knowledge	Importing knowledge from external experts Promoting access to other networks	Hong (2017); Probst & Borzillo (2008); Sanz Martos (2012);
Learning	Access to both internal and external expertise	Hong (2017); Probst & Borzillo (2008); Sanz Martos (2012);
Leadership	Top-management buy-in Advocacy of leaders (champions)	Akhavan et al. (2015); Retna & Ng (2011); Sanz Martos (2012);
Illustrating results and performance	Showing real impact Showcasing smaller goals and sub-objectives Evaluating performance	Hong (2017); Probst & Borzillo (2008); Sanz Martos (2012);
Strategy	Clear, understandable vision Clear, measurable goals and objectives	Akhavan et al. (2015); Hong (2017); Probst & Borzillo (2008);

to learning, which can be supported by getting new inputs from outside the CoP and is related to sharing best practice to not only benefit the CoP, but to also save resources (Probst & Borzillo, 2008). Furthermore, belonging to a CoP can provide significant learning opportunities, both on a personal and group level (Hong, 2017). Co-creation methods that support and foster mutual learning as well as sharing and producing knowledge are those that enable exchange with other experts or practitioners both internal and external (e.g., *Focus Groups*; *Interviews* or *Mentoring Circles*). Sharing best practices can be facilitated by using methods to support the exchange between CoP members (e.g., *Five-Minute Favour*; *1-2-4-All*; *Fish Bowl*).

Development of a CoP as an organised collective is contingent on having a shared mutual culture, values, and a sense of belonging. These aspects are also important when it comes to fostering the other success factors. For instance, to learn is to show one's vulnerabilities, thus a safe, non-judgemental environment is needed, and it requires the community to nurture supportive culture and values around diverse learning needs and deepening knowledge at all levels. These aspects pave the way for knowledge exchange and mutual learning but also interaction and sharing best practice. Thus, mutual culture, values, and belonging form an essential part of the CoP success leading to a deeper interest and commitment to the domain, and consequently to the CoP (Retna & Ng, 2011). Trust, common values, shared understanding, a sense of belonging and cultural awareness influence the level of commitment (Fontainha & Gannon-Leary, 2008).

However, to create a common culture and a sense of belonging, it is crucial to interact and communicate regularly and actively participate in the community (Pyrko, Dörfler, & Eden, 2017) both via face-to-face meetings, and virtually (Fontainha & Gannon-Leary, 2008). Communication and interaction within a CoP can be supported by a facilitator or by providing suitable communication channels (Jagasia et al., 2015). To enable development of a mutual culture and a sense of belonging, we suggest methods that nurture team spirit and social exploration to focus not only on personal objectives and visions but also on strengthening interpersonal relationships between the CoP members (e.g., *Heart, Hand Mind*; *Four Quadrants*).

To support CoPs in defining, as well as maintaining, their culture and values, and facilitating community interaction, leadership is yet another critical factor (Retna & Ng, 2011). Skilful leadership within a CoP can also provide inspiring and engaging blueprint for communication or resource exchange through digital and face-to-face platforms, which at the same time fosters commitment, motivation, and stimulus needed for CoPs to succeed (Akhavan et al., 2015; Retna & Ng, 2011). Through effective leadership, CoPs can also contribute to the success of the hosting institution if the CoP forms an integral part of the broader institutional strategy (Hong, 2017), e.g., incorporating equality, diversity, and inclusion as part of the institutional values and vision. This success factor can be cultivated by using co-creation tools that allow CoP members to seek support from leadership (e.g., *What I*

Need from You; Interviews); or support the CoP leader or facilitator in planning activities (e.g., *Critical Uncertainties*).

Inspired and motivated members play a significant role in fostering a successful CoP, which can be accomplished by communicating achievements and impact (Probst & Borzillo, 2008). Here, evaluations of the CoP can be beneficial as they often lead to increased effectiveness and improved performance (Hong, 2017). To create a platform for reflection on the completed work and to plan next steps, we suggest employing methods that encourage deep questioning (e.g., *What, So What, What Now? Fish Bowl*), and help to illustrate results (e.g., infographics, storyboards, photo documentation).

Closely linked to several of the other success factors is the importance of having a strategy and long-term goals. Akhavan et al. (2015) explain that having clear and concrete strategies lead to an increase in motivation and commitment. Furthermore, defining a strategy and specific goals have a very high impact on the success of a CoP. These processes can be supported by using co-creation methods that help with project planning (e.g., *Plan of Change; Argument Mapping; Future Workshop*).

Arguably all the success factors are interlinked and interdependent, nonetheless, distinguishing the different aspects that make CoP successful is necessary to provide targeted support in the form of specific co-creation activities to assist CoPs in progressing.

Co-creation methods for CoPs

Having outlined the background to the CoP theory, we now provide a selection of four methods described in more depth that are suitable for a workshop for facilitating CoPs both online and face-to-face: *DAKI Retrospective, 1-2-4-All, Plan of Change, and Future Workshop*. Table 4.2 maps each presented method with its respective theoretical link.

Table 4.2 The four presented methods and their theoretical links.

Method	Lifecycle phases	Main areas of activity	Success factors
DAKI Retrospective	Inquire Design Grow Sustain	Learning and developing practice Taking action as a community	Sharing best practice Learning Strategy
1-2-4-All	Inquire Design	Learning and developing practice	Community interaction Sharing best practice
Plan of Change	Design Prototype	Taking action as a community	Leadership Strategy
Future Workshop	Inquire Design Prototype	Learning and developing practice Taking action as a community	Community interaction Mutual culture, values, and belonging Leadership Strategy

DAKI retrospective

Aims of the Method: Retrospective activities (lat. *retrospectare*, "to look back") have been extensively used in agile working frameworks, typically in software development, allowing teams to reflect on work effectiveness and efficiency, to establish continuous improvement processes, and to trigger a cultural change in organisations or traditional change activities (Loeffler, 2011). The key aim of the activity is to gain knowledge and insights from the past processes and determine what needs to be changed.

Retrospectives should not only focus on negative points but explore potential success stories and best practices to learn from. Ideally, retrospectives should be led by a facilitator who is not at the same time also a participant, and the activity needs to include the entire community involved in the project to ensure rich and diverse results (Kerth, 2001). Moreover, the activity must take place in a "safe" environment for the participants to feel secure within their community, which takes time to be developed and maintained. Establishing such feelings of safety will be part of the earlier CoP lifecycle phases, but in the later phases (such as *grow* and *sustain*), trust, honesty, and assurances of no retribution should have already been demonstrated.

DAKI retrospective is a variant of more broadly defined retrospectives that supports the assessment of the status quo, practice, or process based on the four categories of *drop, add, keep,* and *improve. DAKI* is especially helpful when reviewing a task, project, or reflecting on a process to evaluate what should not be continued (*drop*), what could be added to support the project (*add*), what was successful (*keep*), and lastly, what needs further improvement (*improve*).

It works particularly well for nurturing the success factors of *sharing best practice, learning,* and *strategy*; and *learning and developing practice* and *taking action as a community* as the two areas of CoP activity. It is well-placed to be introduced in the *inquire, design, grow,* and *sustain* lifecycle phases.

This activity has proven to be effective in the ACT Matching Events, where it was used in three different breakout sessions on (1) how to compensate for COVID-19 in contract, tenure and promotion decisions in HE and research careers; (2) supporting academics with care responsibilities in working from home during the COVID-19 pandemic; and (3) surveying the experiences and perceptions of researchers on the consequences of the pandemic (see Figure 4.1). The participants enjoyed the interactive method of discussing and sharing ideas virtually, but the results have also been used to draft a report and a summary of possible policy interventions, or potential future steps.

Preparation of the Method: For this method, a matrix drawn on a flipchart is needed or, if the meeting is virtual, by using a (preferably interactive) whiteboard tool. The resulting squares are marked with the four categories of *drop, add, keep,* and *improve*. There is no rigid recommended group size for this activity, however larger groups might need to be divided into smaller circles and a plenary session with all participants at the end. There

DROP...

Drop out of rankings and excellence frameworks...

The rhetoric that this is a great time to finally finish those papers/ monographs

All kinds of faulty metrics

Remove the Keeping in Touch (KIT) Days

Considering caring to be an 'equality issue' when we're talking about workload issues!

ADD...

Shorter working week for same pay and same outputs

This will be controversial. Cap the number of articles counted for promotion. It could help with balancing the caring duties?

Ask for narrative style CVs that focus on career breaks, reasons and personal achievements

Drop the promotions process and introduce small incremental salary scales. Reward maternity and careers leave with an increased increment. Signals support for this activity and no disadvantage. Change attitudes in the long term maybe/hopefully

Online teaching support

Provide a protected research period for the equivalent time that someone was on maternity/carers leave

Cap n of papers considered for promotion etc to no more than e.g. one per year

reduce pressure on reviewers in the publication process

Flexible deadlines

Set up survey to detect where the problems are

KEEP...

Additional resourcing to support transition to online teaching

Keep the EDI roles - especially at VP level

In my Institution: Dropping the research impact/output requirement from the next Appraisal/PDP

Grant prolongation in case of family leave for all funding schemes

Adjust promotions/ probation criteria to allow for impact of caring - but on an opt-in basis

Messaging that it's OK to be doing the best you can

Significant IT support and support for designing online sessions

IMPROVE...

4 day working week:)0

support IT and caring

Offer specific COVID-19 support grants

Offer additional funding to support staff before, during and after parental leave

Improve how funders view and support maternity leave for researchers

Trainings for teaching staff just entering university

Promote life/care activities across all staff, encourage men to 'care' more

Rebuilding Research Momentum Fund - in place but needs monitoring/ adjustment

Improve communication on currently available measures, encourage men to take family leave

Analysis of (potential) equality impacts of new policies and processes

Promote slow scholarship ⟶ Link to the book: https://utorontopress.com/ us/the-slow-professor-3

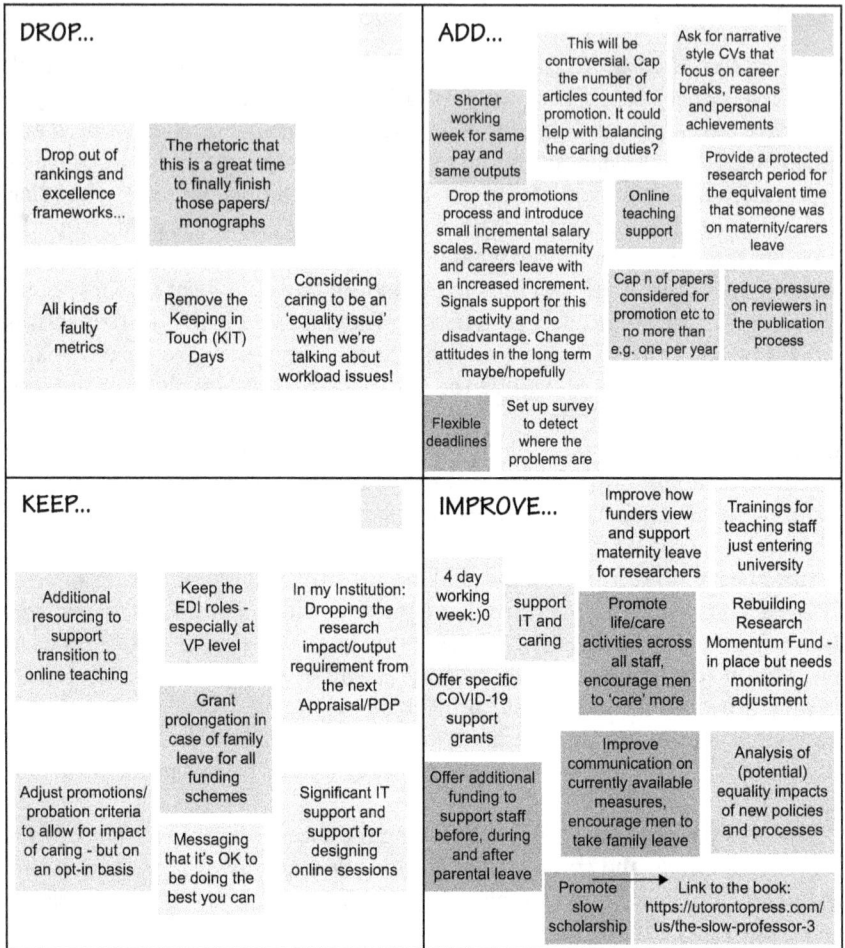

Figure 4.1 DAKI wall outputs at ACT matching events hosted by Science Foundation Ireland.

is not much pre-work required from the facilitator, however, it is advised to familiarise oneself with the community members and the project to review (Kerth, 2001).

Structure of the Method: The participants are invited to reflect on the previously agreed topic and identify at least one entry for all four variables and write them down on post-it notes or cards. After shortly explaining the content by the participants, these cards will be pinned on the flipchart or pinboard in the relevant square (*drop, add, keep,* and *improve*). Similar comments can be merged to avoid unnecessary repetitions. If facilitating this activity online, the participants can write their comments independently in the matrix on the shared whiteboard or have them collated by the facilitator who applies all contributions onto the virtual whiteboard. The process can be repeated several times during the same session with different issues, topics, or projects for review.

1-2-4-All

Aims of the Method: Methods that divide large groups into smaller discussion subgroups such as *1-2-4-All* (also known as buzz groups, snowballs, or pyramids) aim to foster engagement of every individual and support exploring the existing knowledge both at an individual and group level. These methods are often applied in education (Romeike & Fischer, 2019) and are commonly used to start a discussion, to generate new ideas, or to reach consensus. Through the process of "thinking together" (Pyrko, Dörfler, & Eden, 2017), this method offers CoPs, the structure to facilitate problem solving, and it can be used in combination with other techniques (Brewer, 1997).

The *1-2-4-All* activity supports the development of the success factors for *community interaction* and *sharing best practice* (and *learning and developing practice* as one of the main areas of CoP activity), but it can also support CoPs in the first two lifecycle phases (*inquire, design*).

Whilst this method was also tested in the ACT CoP facilitator training, it has been used frequently by different ACT CoPs as well. The GenBUDGET CoP used *1-2-4-All* in their (virtual) consolidation workshops, that took place in April and May 2020. This CoP used the activity more than once with virtual breakout rooms with success resulting in fruitful conversations and interactive discussions.[1]

Preparation of the Method[2]: This method does not require any material except for a flipchart and pens (or virtual whiteboard) should the discussion results be recorded. The facilitator or the group decide on the focal issue before the session. Questions to guide the course of the activity could be: *What ideas/ways and possibilities do YOU identify/have in mind to support/ progress the issue at hand?* There is no limit on how many people can take part in this activity. However, if the group is large, it is possible to adjust it by increasing the number of participants after the individual reflection stage (e.g., *1-3-6-All*). This activity can be adapted to online workshops using breakout sessions.

Structure of the Method[3]: The activity starts with quiet reflection where each participant individually (*1*) brainstorms the presented issue or question. Next, participants pair up (*2*) to exchange their ideas, and then discuss the issue in groups of four (*4*). It might be helpful for participants to write down the ideas, agreements, disagreements, and problems to share in the next step of the activity. Lastly, the method ends with everyone (*All*) discussing the results in in the plenary to identify the most prominent or illuminating ideas. Each group can nominate a spokesperson to present their group's outputs.

Plan of change

Aims of the Method: Plan of Change is an example of a method broadly based on the theory of change, "an outcome-based approach which applies critical thinking to the design, implementation and evaluation of initiatives and programmes intended to support change" (Vogel, 2012, p. 3). A critical lens is applied to the surrounding conditions that influence the plan, the motivations and contributions of stakeholders and other actors, and the different interpretations or assumptions in relation to how and why that order of change might arise. The method is simply "the description of a sequence of events that is expected to lead to a particular desired outcome" (Davies, 2012, para. 6), and it is particularly helpful in making assumptions explicit. These assumptions act as "rules of thumb" and they reflect deeply held values, norms, and ideological standpoints. Making these assumptions obvious, and critically reflecting how they can influence our choices, theory of change encourages constant "questioning of what might influence change in the context and drawing on evidence and learning during implementation" (Vogel, 2012, p. 4). The method was effectively used at the start of the ACT project during a face-to-face facilitator training event to test the method's useability and group dynamics in a safe and informal environment.

The *Plan of Change* activity supports CoPs in *leadership* and *strategy* as the success factors; *taking action as a community* as one of the areas of CoP activity; and it can also support CoPs in the *design* and *prototype* lifecycle phases.

This method has been used in the ACT CoP facilitator training. Our experience has shown that it is crucial to mutually agree what each step connotates to ensure that all participants operate with the same understanding of the categories. This avoids confusion and clarifies the nuanced meanings of the five-step sequence (Table 4.3).

Table 4.3 The five-step sequence categories for the *Plan of Change* method.

Long-term impact	The "big picture", the ideal scenario, often quite speculative. Can be based on the belief that some *short-term impacts* and *results* may lead to broader applications or development, e.g., achieving gender parity among professors in the school of science by 2025.
Short-term impact	The development of a temporary or partial solution to the "bigger issue"; e.g., facilitating access to gender disaggregated data on promotions for monitoring.
Results	The outputs, the products of the *activities,* e.g., gender equality status-quo reports, staff survey results.
Activities	Actions, instructions, contact points, control visits, e.g., creating gender disaggregated data for reports, regular survey administration, evaluation, action plans.
Resources	Budgetary, human resources, information and communication technology, access to expertise, access to institutional decision making, e.g., access to training for survey administration, staff, expertise in data collection, etc.

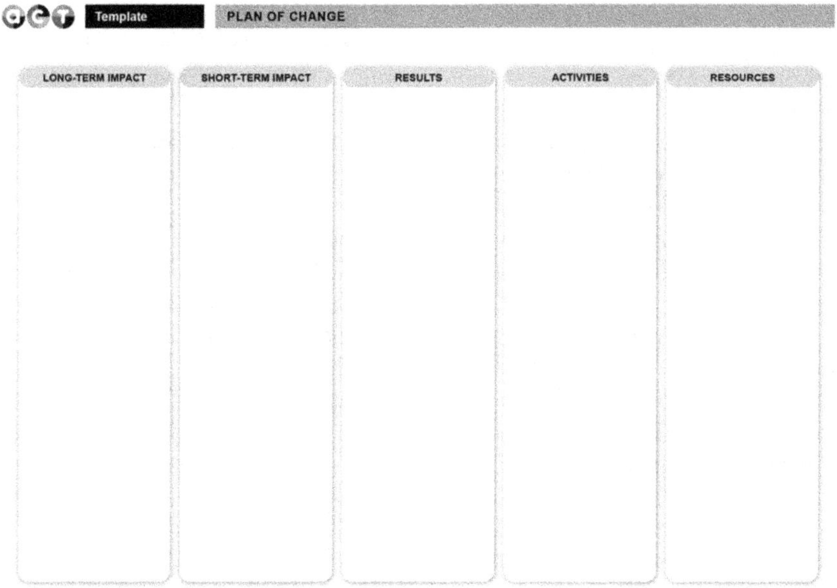

Figure 4.2 Plan of change template.

Preparation of the Method: Plan of Change requires time, as it is a multi-stage process, from shorter 2–4 hours sessions for rapid planning (face-to-face or online), or whole day workshops for more complex projects. The participants should possess the necessary breadth and depth of knowledge on the topic and context.

Structure of the Method: An essential part of this activity is the provided template with the five-step sequence (*long-term impact*; *short-term impact*; *results*; *activities*; *resources;* Table 4.3), which can be printed or transferred to a flipchart or a virtual whiteboard.

The provided template (see Figure 4.2) guides the activity from *impacts* via *results* and *activities* towards *resources*. This can also be done vice versa, or iteratively. Pathways to *impacts* and *results* should be drawn. Lastly, critical assumptions on which the plan is based should then be identified and listed: *What questions does the change plan pose, and how can new activities and resources support the work to achieve the aimed effects?*

Future workshop

Aims of the Method: The origin of *Future Workshop* can be traced to the early work of Jungk and Müllert (1987), who developed this method for the purpose of enhancing democratic municipal decision making by allowing the residents an opportunity to influence the future of their town. This method would later become a tool in the political fight of civil action groups for a better enforcement of their interests, and thus creating a desirable future. Jungk and Müllert's objective was to facilitate the development of "social fantasy" leading to conflict resolution and turning the critical citizen into an emancipated subject (Apel, 2004).

Capturing the philosophy of Critical Theory and the Hegelian dialectics, the underlying principle of the method is that problems can only be solved by critique to expose and unmask the status quo (Apel, 2004). Thus, the method is oriented towards collective efforts for institutional and political change. Following this logic, *Future Workshop* starts with an explicit critique phase, and in the subsequent fantasy phase the critique outputs provide a springboard for reimagining the future. The method is a creative and playful way for participants to design their desired, ideal future without restricting imagination e.g., by refraining from thinking solely about financial resources. Thus, the aim of this activity is two-fold: first, to design a desired future; and second, to develop unconventional and imaginative solutions to an issue or a problem. Overall, *Future Workshop* is most effective with an atmosphere that promotes creativity and visionary thinking, as it leads to new perspectives and a clear view of future developments and possibilities for oneself and the institution. This activity can be run as a standalone activity, but also works well in combination with other methods, e.g., the outputs from other activities can be used as a stimulus for this method.

The *Future Workshop* activity supports the development of the success factors for *community interaction; mutual culture, values, belonging; leadership*, and *strategy* (and *learning and developing practice;* and *taking action as a community,* as two of the main areas of CoP activity), but it can also support CoPs in the first three lifecycle phases (*inquire, design,* and *prototype*).

This method has been effective in the CoP facilitator training, where the group was invited to draw their imagined futures on large sheets of paper to create a gallery of diverse CoP visions, or alternatively to compose prose or a poem. It fostered emergence of participants' interpersonal relationships through producing and presenting large-sheet drawings from the fantasy phase pinned to the walls. Additionally, the three phases of *criticism, vision,* and *implementation* were drawn out on the floor (using adhesive tape) to represent movement and include a spatial dimension to the understanding of the phases as a long-term process and a "journey" that requires space, time, and resources. The activity was very well received and energised the participants with creativity and movement.

Preparation of the Method: The success of the method relies on honest and active participation, which is why it is important to foster a collaborative, relaxed, and informal environment. The design of this method can incorporate highly creative outputs, i.e., participants might be encouraged to produce poems, narratives, drawings, sculptures, dance, drama, etc. Accordingly, various props might be needed, however ordinarily, flipchart, large sheets of paper and pens will suffice, or virtual whiteboards for online workshops will be required.

Structure of the Method: This activity consists of three phases: the *criticism* phase, the *vision* phase, and the *implementation* phase. Depending on the group dynamic, the order of the phases can also be modified (i.e., start with the *vision* phase) as to not demotivate and lock participants into negative thinking as they begin. However, it is important to note that honest critique is paramount to this method, and as such, this adaptation should only take place if there are noteworthy reasons for this, i.e., broaching sensitive topics, very early phases of the CoP lifecycle, working with members who have not yet forged close relationships, or with communities who are still in the process of establishing trust and openness.

In the first phase, participants reflect on the status quo and their points of critique. In the second phase, participants are encouraged to envision their desired future. At this stage, obstacles that might occur in this utopian vision are consciously ignored, which allows participants to think outside of what is practicable or realistic. Thus, an inspiring vision for future success can be created. The implementation phase is used to structure the ideas and suggestions collected. Participants can then evaluate if these ideas are realistic and viable (e.g., *What needs to happen if the vision is to be delivered?*) and plan the next steps.

Conclusion

In this chapter, we have presented a selection of co-creation methods aimed at facilitating CoPs in the efforts for institutional change. In doing so, we address the gap in the literature with respect to the practical tools and methods for working with CoPs. As Pyrko Dörfler, & Eden (2017) suggest, CoP members need to think more intentionally about the three structural CoP elements (*domain, practice, community*). We add to this vein of thought and further suggest that CoP facilitators and members should be inspired to design their community interactions mindful of the CoP lifecycle phases, success factors, and the main areas of activity through which their communities mutually engage and "think together" (Pyrko Dörfler, & Eden 2017).

The four methods detailed in this chapter allow the CoPs to foster a critical appraisal of the status quo, exploration of one another's views, engagement in productive conversations, planning of next steps, and eventually promoting and accelerating institutional change. The reason for selecting these methods is that they are well-placed to address the most significant needs of CoPs embarking on equality and institutional change projects: they need to be inclusive, allow diversity of perspectives, and lastly facilitate collaborative imagination of the future and the planning of change, which often requires a courageous vision.

We recommend that future research applies the introduced methods when working with CoPs for institutional change, and in this way, helps to establish the effectiveness of the methods towards change programmes, explore the challenges of various aspects of facilitation and participation, online adaptations, and group and power dynamics.

Notes

1. This information has been provided by the GenBUDGET facilitator (Laufey Axelsdóttir) and the GenBUDGET blog: https://genbudget.act-on-gender.eu/Blog/genbudget-consolidation-workshop.
2. Adapted from: https://www.liberatingstructures.com/1-1-2-4-all/.
3. Ibid.

References

Acker, Joan (2000). Gendered contradictions in organizational equity projects. *Organization*, 7(4), 625–32. doi.org/10.1177/135050840074007

Akhavan, Peyman, Marzieh, Babaeianpour, & Mirjafari, Masoumeh (2015). Identifying the success factors of Communities of Practice (CoPs): How do they affect on students to create knowledge? *Vine*, 45(2), 198–213.

Apel, Heino (2004). *The Future Workshop*. Deutsches Institut für Erwachsenen-bildung. Retrieved from: www.die-bonn.de/esprid/dokumente/doc-2004/apel04_02.pdf [20.05.2021]

Axelsdóttir, Laufey (2020). *GenBUDGET Consolidation Workshop*. Available from: https://genbudget.act-on-gender.eu/Blog/genbudget-consolidation-workshop.

Brewer, Ernest W. (1997). *13 Proven Ways to Get Your Message Across: The Essential Reference for Teachers, Trainers, Presenters, and Speakers*. California: Corwin Press, Inc.

Cambridge, Darren, Kaplan, Soren, & Suter, Vicki (2005). Community of practice design guide: A step-by-step guide for designing & cultivating communities of practice in higher education. *EDUCASE Learning Initiative (ELI)*.

Davies, Rick (2012). Criteria for assessing the evaluability of a theory of change. Retrieved from: http://mandenews.blogspot.co.uk/2012/04/criteria-for-assessing-evaluablity-of.html. [13.05.2021]

Fontainha, Elsa, & Gannon-Leary, Pat (2008). Communities of practice and virtual learning communities: benefits, barriers and success factors. *E-learning Papers*, 5, 20–9.

Frow, Pennie, Nenonen, Suvi, Payne, Adrian, & Storbacka, Kaj (2015). Managing co-creation design: a strategic approach to innovation. *British Journal of Management*, 26(3), 463–83.

Griffith, Terri, Fuller, Mark A., & Northcraft, Gregory B. (1998). Facilitator influence in group support systems: Intended and unintended effects. *Information Systems Research*, 9(1), 20–36. Retrieved from http://www.jstor.org/stable/23010868. [06.07.2021]

Hogan, Christine (2002). *Understanding Facilitation: Theory and Principles*. London: Kogan Page.

Hong, Jongyi (2017). A method for identifying the critical success factors of CoP based on performance evaluation. *Knowledge Management Research & Practice*, 15(4), 572–93.

Jagasia, Jyoti, Baul, Utpal, & Mallik, Debasis (2015). A framework for communities of practice in learning organizations. *Business Perspectives and Research*, 3(1), 1–20.

Jungk, Robert & Müllert, Norbert (1987). *Future Workshops: How to Create Desirable Futures.* London: Institute for Social Inventions.

Kerth, Norman L. (2001). *Project Retrospectives: A Handbook for Team Reviews.* New York: Dorset House Publishing Co, Inc.

Loeffler, Marc (2011). *Improving Agile Retrospectives: Helping Teams Become More Efficient.* Pearson Education, the United States of America.

McDermott, Richard (2000). Community development as a natural step: Five stages of community development. *Knowledge Management Review*, 3, 16–9.

Osborne, Stephen, & Strokosch, Kirsty (2013). It takes two to tango? Understanding the co-production of public services by integrating the services management and public administration perspectives, *British Journal of Management*, 24 (S1). doi.org/10.1111/1467-8551.12010

Probst, Gilbert, & Borzillo, Stefano (2008). Why communities of practice succeed and why they fail. *European Management Journal*, 26(5), 335–47. doi.org/10.1016/j.emj.2008.05.003

Pyrko, Igor, Dörfler, Viktor, & Eden, Colin (2017). Thinking together: What makes Communities of Practice work? *Human Relations*, 70(4), 389–409. doi.org/10.1177/0018726716661040

Pyrko, Igor, Eden, Colin, Dörfler, Viktor, & Stierand, Marc (2017). *Facilitating communities of practice with causal mapping workshops* [Proceedings Paper]. Ifkad 2017: 12th International Forum on Knowledge Asset Dynamics: Knowledge Management in the 21st Century: Resilience, Creativity and Co-Creation, 1966-77.

Retna, Kala S., & Ng, Pak Tee (2011). Communities of practice: Dynamics and success factors. *Leadership & Organization Development Journal*, 32(1), 41–59.

Romeike, Bernd F. M., & Fischer, Martin (2019). Buzz groups facilitate collaborative learning and improve histopathological competencies of students. *Clinical Neuropathology*, 38(6), 285–93. doi.org/10.5414/np301195

Sanz Martos, Sandra (2012). *Comunidades de Práctica: El Valor de Aprender de Los Pares.* Barcelona, Editorial UOC.

Thomson, Aleksandra, & Rabsch, Kathrin (2021). *ACT - Community of Practice Co-creation Toolkit (2.0).* Zenodo. doi.org/10.5281/zenodo.5342489

van den Brink, Marieke (2020). "Reinventing the wheel over and over again". Organizational learning, memory and forgetting in doing diversity work. *Equality Diversity and Inclusion*, 39(4), 379–93. doi.org/10.1108/edi-10-2019-0249

Vogel, Isabel (2012). *Review of the use of 'theory of change' in international development. Review report.* The UK Department of International Development. Retrieved from https://www.theoryofchange.org/pdf/DFID_ToC_Review_VogelV7.pdf [19.05.2022].

Wenger, Etienne, McDermott, Richard A., & Snyder, William (2002). *Cultivating Communities of Practice: A Guide to Managing Knowledge.* Harvard Business School Press, Boston.

5 Communities of Practice and gender equality

Fostering structural change in research and academia in Central and Eastern Europe

Paulina Sekuła, Ewelina Ciaputa,
Marta Warat, Ewa Krzaklewska,
Sarah Beranek and Sybille Reidl

Introduction

In recent years, gender equality in higher education (HE) and research and innovation (R&I) has been high on the European policy agenda. The European Commission supports research performing organisations (RPOs) and higher education institutions (HEIs) in introducing gender equality measures, including comprehensive gender equality plans (GEPs). Despite some gains (Timmers et al. 2010; European Commission 2019), the implementation of structural change and reaching sustainable outcomes remain difficult (Cavaghan 2017). Common problems include resistance within institutions, lack of management support, an absence of sustained financial and human resources, unavailability of gender expertise, as well the lack of authority of the staff responsible for developing and introducing GEPs (EIGE 2016). These problems are also witnessed in Central and Eastern Europe (CEE), but additionally some regional specificities play a role. Although anti-discrimination and gender equality legal framework is present throughout the region and includes constitutional and labour code provisions and equal treatment legislation (Böök et al. 2021), specific laws on gender equality in HE and R&I and policy incentives for implementing gender equality measures have not been put in place (EIGE 2016). As a result, the extent of the adoption of GEPs in HEIs and RPOs – perceived as an effective tool for institutional change – is in CEE significantly lower than in other regions of Europe (European Commission 2019; Reidl et al., 2019). Moreover, the institutions in CEE are experiencing stiff political resistance to gender equality interventions and feminist agendas as a part and parcel of democratic backsliding in the region (Krizsán & Roggeband 2019). They are nevertheless developing some internally-driven

DOI: 10.4324/9781003225546-5

initiatives, including the creation of anti-discrimination and gender equality bodies, developing anti-harassment policy and introducing work-life balance measures. More comprehensive gender equality interventions – in the form of implementing GEPs – have been undertaken mainly within the framework and lifespan of EU-funded institutional change projects. While these projects offer a transfer of good practices from countries with higher level of uptake of structural change to those who are "lagging behind", they often lack a sustainable support from intra- and inter-organisational stakeholders. Based mainly on West European experiences, they also risk not taking fully into account the local legal, political, and historical contexts of the CEE region. Moreover, these interventions remain scattered, and there is insufficient flow of information and exchange of experiences concerning them.

The reflection on CoPs stems from the assumption that they are the experimental environment, where alternative practices can be enacted, incorporated into the organisational environment and, thus, slowly accelerate change (Chapter 1, Müller & Palmén, this volume, p. 14). While there are several structural and processual factors that facilitate or hinder the effectiveness of institutional change for gender equality in HE and R&I, we will analyse which of them can be created or fostered by an inter-organisational CoP operating in conditions of lack of coordination of dispersed and isolated activities, insufficient legal and policy incentives, and high levels of resistance. We argue that an inter-organisational CoP operating within the region of CEE may play a role of intermediary support structures that connect various initiatives and strengthen those conditions necessary for structural change through enhancing the capacity and agency of organisational change actors. Such structures may as well allow for exchange of localised, context-specific knowledge and discuss tailored strategies that are possible in the region or national context.

To discuss the role of CoPs in the process of institutional change in HEIs, the chapter analyses, through the case study approach, the process of setting up and developing the CoP for Gender Equality in Central and Eastern Europe (GEinCEE CoP). The CoP was created within the H2020 ACT project and it aimed to respond to particular needs of gender equality practitioners and scholars in the CEE region voiced, among others, in the ACT Community Mapping survey. According to the survey, the most urgent needs include support from organisation management, gender know-how, regular monitoring of gender equality status quo, strategies to overcome resistance and financial resources (Warat et al. 2019).

The GEinCEE CoP's mission is to promote and support institutional change to advance gender equality in HEIs and RPOs, i.e., diagnosing the situation as well as promoting and assuring support in designing, introducing, and monitoring selected gender equality measures or GEPs. It also collects and systematises existing experiences and knowledge in the region. As of June 2021, GEinCEE CoP gathers researchers and gender equality officials of 20 organisations from Poland, Lithuania, Czech Republic, Slovenia, Ukraine, and Cyprus.

In this chapter, we will first present the theoretical considerations in regards to the role of CoPs in supporting institutional change towards gender equality in HEIs and RPOs. After presenting the case-study approach, we will discuss the creation and development of GEinCEE CoP. Finally, we will analyse the possible impact of the CoP on factors supporting gender equality institutionalisation.

CoP approach to structural change towards gender equality

The academic reflection points to several factors that facilitate the effectiveness of institutional change for gender equality in HE and R&I. A single, most important structural element is the conducive governance framework with legally-binding measures and positive incentives for introducing comprehensive gender equality measures (Linková et al. 2007; Palmén & Kalpazidou Schmidt 2019). When this condition is lacking, the fate of gender equality interventions depends on other factors that have been identified across various studies. The key processual factor is the involvement of all organisational stakeholders, including governing bodies and other actors across the whole organisation early in the process (Lipinsky 2014; Vinkenburg 2017; Palmén & Kalpazidou Schmidt 2019). At the same time, resistance to gender equality interventions at different levels and from different actors has been identified as an important obstacle to successful implementation of gender equality interventions (Verge et al. 2018; Palmén & Kalpazidou Schmidt 2019). In this context, framing synergies with other initiatives within an organisation and linking gender equality issues to wider concepts such as research excellence or responsible R&Is is considered as a pro-active strategy to tackle resistance at different levels and, therefore, can be considered a factor enhancing the effectiveness of gender equality initiatives (Colizzi et al. 2019; Palmén & Kalpazidou Schmidt 2019). Similarly, sufficient resources, including funding and decisive power of gender equality bodies are important facilitators of effective gender equality interventions, as well as gender expertise, awareness and competence within organisations (EIGE 2016; Lipinsky 2014; Lansu et al. 2019; Palmén & Kalpazidou Schmidt 2019). They all enhance the capacity and agency of gender equality actors to initiate and sustain institutional change. Any transformations should also be rooted in the organisational aims and structures to proof them from personal changes and enable them to become sustainable (Colizzi et al. 2019). Formulating realistic targets tailored to the context of a given organisation, as well as comprehensive monitoring of the effects of gender equality interventions are deemed to enhance the obligation of the organisational leadership and other stakeholders to actively promote gender equality (Mühlenbruch & Jochimsen 2013; Palmén & Kalpazidou Schmidt 2019). Thus, difficulties in obtaining organisational sex and gender disaggregated data hamper the processes of designing, implementing, and monitoring gender equality interventions (Palmén & Kalpazidou Schmidt 2019).

Previous knowledge, although limited, suggests that a CoP can help create some of the above-mentioned conditions for effective implementation of gender equality interventions. Firstly, by engaging different stakeholders, i.e., gender equality practitioners, researchers, human resources, and top management, CoPs bring together a range of perspectives on a problem and different types of competences, and therefore ensures that relevant and context-specific knowledge is accessible to those who need it (Hearn & White 2009). Secondly by engaging different functional roles in an organisation, CoPs enable the transcendence of organisational hierarchies and functional boundaries and may therefore assure that implementing change is a multi-stakeholder responsibility (Palmén et al. 2019). Thirdly, through emphasising community engagement and participation, CoPs may effectively tackle resistance (Palmén et al. 2019). Fourthly, while this has not been explicitly tested, CoPs may empower their members to pursue and sustain change at their organisations. As collective agency emerges through a learning process – occurring at group discussions, community meetings, participatory workshops or informal conversations (Pelenc et al. 2015) – CoPs seem to be suitable for improving the agency of gender actors. Additionally, by fostering the sharing of practice, mutual learning, and the promotion of the achievements of the gender equality projects that overcome national and institutional boundaries, inter-organisational CoPs contribute to the advancement of gender equality in R&I and HE at the European level (Palmén et al. 2019). With the case study of GEinCEE CoP, we investigate whether and to what extent fostering conditions for institutional change may be performed by an inter-organisational CoP.

Methods

The research draws on the strengths of the case study approach, particularly its heuristic potential, the triangulation of methods and data, and the ability to stress the case evolution in time through a series of interrelated events (Flyvbjerg 2011; Yin 2014). Case study research is an in-depth, detailed exploration of the individual, group or a phenomenon, also a project, a programme, or an institution, aiming at a comprehensive and rich description of an individual case and its analysis (Flyvbjerg 2011; Starman 2013). The case study approach enables a focus on the contextual factors that are relevant to the phenomena being studied (Yin 2014), or the "relation to environment" (Flyvbjerg 2011), which is important to highlight specific conditions for the community in the CEE region.

The empirical evidence for this study is of a secondary character and based on the documentation of setting up and developing the GEinCEE CoP from December 2018 until June 2021. The analysed data come from the ACT project evaluation process, which included the aim of identifying challenges and strategies for developing CoPs for institutional change, assess the usefulness of the learning outcomes for the CoPs, as well as the effect of the CoPs on the development of gender equality in their member institutions. Progress reports were written by the CoP facilitators that document the development of the CoPs based on the routine filling in of the monitoring files. They contain detailed information on the CoPs' objectives, composition, activities and progress, as well as meeting protocols, social media content such as blogposts or tweets. Additionally, within the evaluation process 3 semi-structured interviews with CoP members were conducted online in 2020 by the ACT project partner JOANNEUM RESEARCH. The interviewees were selected following criteria that aimed at getting a diverse picture (e.g., in terms of their organisation size, country, or number of CoP meetings attended). The interview questions targeted participation in CoP activities, cooperation and communication with other members and perspectives on the benefits and impact of CoP involvement. Furthermore, they concerned the perceived limitations of the CoP approach and further needs to achieve structural change in member organisations and in the European Research Area.

Beyond the ACT project evaluation data, the chapter also draws from analysis of written summaries and any other documentation from 12 general CoP meetings, two CoP member-only workshops, and three open events. The summaries from all events between 02.2019 and 05.2021 report meeting topics and the discussed issues. Also, the content of the blog run by the members of the CoP was analysed. As of June 2021, it includes 27 posts prepared by both GEinCEE CoP's members and CoP co-facilitators.[1]

Additionally, to provide a context for the emergence of the GEinCEE CoP, selected results from the ACT Community Mapping survey were used. The survey – carried out in 2018 – mainly reached people involved in the processes of implementing gender equality measures in HEIs and research institutes. Its aim was also to identify potential community members and their needs (Reidl et al. 2019).

Finally, the individual experience of the chapter authors concerning participation in the GEinCEE CoP is used as data. Two authors, Paulina Sekuła and Ewelina Ciaputa, serve as CoP facilitators and coordinators of CoP working groups. Ewa Krzaklewska is a local coordinator of the ACT project and a researcher. Marta Warat is a researcher in the ACT project and the coordinator of one of the working groups of the CoP.

Built from scratch: On developing and consolidating a Community of Practice for Gender Equality in Central and Eastern Europe

Creation of GEinCEE Community of Practice

The creation of GEinCEE CoP started in December 2018. Initially, Jagiellonian University in Kraków (JU), which was responsible for CoP creation and facilitation, considered focusing primarily on Polish institutions and used pre-existing (in)formal networks of collaboration in the field of gender equality to start its development. After several meetings among Polish prospective members, it was decided that the CoP's reach should be expanded. Individual researchers from the CEE region who took part in H2020 gender equality projects and respondents who declared their interest in joining a CoP in the ACT Community Mapping Survey were invited to several meetings for advice in relation to CoP creation, its potential aims and regional focus. Building on the result of those meetings as well as recently forged bonds between experts, the CoP was created in May 2019 and its name, mission, vision and agenda were determined.

While different types of competences were brought together by the engagement of diverse stakeholders (Hearn & White 2009) including representatives of universities and research institutes, the discussions with the members-to-be and the results of the Community Mapping reinforced the belief that the regional focus is important as countries within CEE share similar aims, concerns, needs, and institutional context. Indeed, the regional focus of the CoP is seen as its main strength, as underlined by the members. Focusing on the CEE region is beneficial in terms of knowledge sharing and providing a sense of belonging:

> I think it is important that this community of practice is focused mainly on this region. And it means, at least for me, that I am part of some network. I can always write to [listing names] asking about information, help, advice, some materials ...
>
> *(Member 3, interview)*

The geographical aspects appear critical in CoP development analysis. Aside from being helpful for developing a cross-national network of gender experts, the CoP also proved critical in terms of strengthening national networks of two of the biggest groups of Polish and Lithuanian experts. The regional focus paradoxically triggered the twinning of organisations from the same country and intensified national-level communication between those organisations and fostered the sharing of practices. Although the CoP has been gaining new members and supporters, it is far from representing most countries of the region and its membership is skewed towards Polish HEIs. Despite the high efforts to engage national policy makers and research funders, these aims to date have not been reached. These problems are perceived by members as weakening the possibility to impact wider national and regional contexts.

Building mutual engagement, joint enterprise, and shared repertoire of action

The creation of the CoP institutionalised collaboration between different types of organisations and individuals operating in the CEE region. The

Memorandum of Understanding (MoU) – an agreement signed by the GEinCEE CoP members – provided a formal framework for cooperation and confirmed members' engagement in the activities of the CoP aiming at advancing gender equality. The CoP won acceptance from central management of most organisations represented in the CoP as only in five cases the MoU was signed by individuals, not the legal representatives.

Members of the CoP took part in several face-to-face meetings, participated in international conferences and debates, which allowed them to meet in person also on an informal basis. It is important to point out that it was face-to-face meetings at the start of the CoP development that constituted a firm base for the community's further operation, as these were considered more valuable when it came to creating relationships and having discussions compared to online meetings. Nevertheless, after the emergence of the pandemic caused by COVID-19, the CoP members took part in over a dozen online meetings (of both the whole CoP and working groups), as well as in tailored workshops and trainings. The CoP had also been present in social media and in scientific discourse through the active dissemination of news, activities and blog posts on the GEinCEE webpage and Twitter, but also online campaigns and discussions organised within the ACT project.[2]

CoP meetings, workshops, and trainings allowed for collaborative learning and knowledge sharing through exchange of good practices. At each CoP meeting, selected members presented developments and challenges to gender equality at their institution. As showed below, the exchange of knowledge supports institutional change, but also hearing other experts' stories creates a sense of belonging and raises motivation for action:

> Therefore, participation in the CoP activities helps us keep a critical attitude towards existing organisational practices and procedures, approaching them from gender equality perspective. More importantly, this participation facilitates our search for the most efficient future steps developing strategies for initiating institutional changes in the most optimal way. In addition, the sense of community and belonging, and the empathic understanding of the complexity of endeavours to strengthen gender equality give strength and motivation to continue striving for better work and life conditions for everyone at Vilnius University.
>
> *(Blog entry 1)*

Participation in meetings, discussions, and cooperation on common problems also allowed the CoP members to overcome isolation, which would otherwise be hindered:

> For me, it was important to experience and get to know more deeply that we are not alone. There are also other institutions ... that face the

same or even worse problems....we can talk and brainstorm about it. We can try to find solutions together. Not to mention the kind of knowledge about what is happening in my region in terms of gender equality....

(Member 2, interview)

To engage in academic discussion around gender equality interventions, the CoP co-organised an international conference and a panel debate that raised a discussion among the most important GE actors in the region, including many from Horizon 2020 structural change projects. These initiatives made their voices heard in relation to the problem of the sustainability of measures and policies beyond the duration of these projects.

Evolution of the GEinCEE CoP – Enhancing visibility and supporting sustainability

After about one year of functioning, the GEinCEE CoP started to evolve from being mainly a forum to exchange knowledge, good practices, and emotional support to an agent with a sustainability and impact strategy. Initially, CoP facilitators focused mainly on group building and providing members with specific tools (discussed in the next chapter) dedicated to reinforcing change in organisations. After about 1.5 years, their focus switched to raising visibility of the CoP and gender equality issues in a wider social context, as well as community sustainability beyond the ACT Project. The Consolidation Event in April 2020 was an important step in this process. During it, members reflected on the status quo of the CoP, indicating its strengths and weaknesses, and picturing its future development. Gaining visibility of the CoP was seen as a decisive factor that triggers interest from other entities potentially joining the CoP, but it also raises the CoP profile, in turn legitimising the activities at member institutions. Therefore, to increase the CoP's visibility and impact, CoP members decided to create three working groups focusing on the preparation of a policy brief (basing on the results from the GEAM survey), an edited book on gender equality in the CEE region and the CoP's sustainability after the ACT project lifespan. Moreover, in order to advance gender equality within member institutions, the CoP Support Programme was set up. The idea was to facilitate change by providing institutions with services indicated by them as needed: support in analysis of the GEAM survey results, organisation and provision of trainings as well as the proofreading of scientific articles on gender equality.

From June 2021, CoP members have been working intensively on the CoP's sustainability plan. They search for the best solutions to maintain the CoP's activities, members involvement and organisational leaders' support. They are concentrating on preparing a proposal for the Horizon Europe call focused on gender equality plan development and support.

How the GEinCEE CoP fosters (or not) structural change

After looking at the process of CoP creation and development, we will analyse to what extent the GEinCEE CoP has impacted on factors facilitating and hindering effective gender equality interventions in its member organisations and the region. We will reconstruct the aspects of this impact according to the order of significance attributed by CoP members.

Agency of gender actors and engagement of organisational stakeholders

The activities developed within the GEinCEE CoP were recognised by their members as effective methods of strengthening the capacities of gender equality supporters to advocate for change in their organisations, as well as supporting the process of institutionalisation of gender equality in member organisations.

> In general, participation in the CoP's activities encourages us to look critically at and react towards all institutional procedures and measures where (potentially) inequality could be practiced. Thus, joining international CoP inspires us to act for institutional change – to clarify and specify institutional gender equality policy, elaborate reasonable and, believable, innovative GEP and, while implementing it, to strive for the elimination of gender bias and all possible inequality supporting practices from institutional procedures by making them transparent and gender sensitive.
>
> *(Blog entry 2)*

Being part of a network of committed individuals and experts, and participation in exchange of knowledge, experience, and practices, was perceived as not only providing a sense of belonging (Pelenc et al. 2015), but also giving expertise and courage to effectively lobby organisational officials for structural change:

> … I succeeded in raising interest in some people also from the managerial board and from the director's circle. And they are ready to implement some plans with me. So, I think it's a big success at that level.
>
> *(Member 1, interview)*

Some CoP members were also able to engage their co-workers in the discussions about possibilities to implement gender equality measures and to involve them in action groups with concrete tasks to fulfil:

> … It was the day when a meeting of the GEinCEE CoP has been organised at the LSRC. This event was a stepping-stone for us: after the workshop, an initiative gender equality group of expert researchers and those

interested in gender equality issues came together and started planning and implementing such activities as the application of the GEAR tool for the assessment of gender equality status quo in LSRC, the participation in preparing an application for H2020 Swafs' program, the organisation of an international conference focused on discussing the implementation of GEPs in RPOs and RFOs in CEE countries, etc. The GEinCEE CoP not only triggered those activities at LRSC, but also took and still is taking active part in most of them.

(Blog entry 2)

In this context, it was pointed out that participation of an organisation in an international network brings the topic of gender equality into the discussion. Organisational leaders become aware that gender equality is undergoing important developments in other institutions in the region that are part of the community. In order not to lag behind, they commit to advance gender equality by supporting or engaging in similar developments at their organisation. However, CoP members were not able in every organisation to get sustainable leadership support and engage more people to join the CoP and to share work related to gender equality. The reasons reported by members were manifold and included lack of time and financial resources, discontinuity due to changes of university governments as well as resistance towards gender equality interventions (which will be discussed in more detail below).

Gender know-how

Previous research highlights gender expertise, as well as practical competences and experience as facilitating factors in implementing gender equality interventions (Palmén & Kalpazidou Schmidt 2019). This has also been widely recognised by the CoP members as a key condition for structural change. Since the beginning of its operation, the GEinCEE CoP has been strongly focused on the exchange of knowledge and practices between CoP members and on knowledge acquirement from external gender experts. CoP members recognise the specificity of the CEE region and appreciate that they can share context-dependent and experience-based knowledge, which can serve as an inspiration to develop more effective tools and measures. Additionally, the CoP makes CoP members' voices heard in the discussions on gender equality which, as expressed during meetings and workshops, to a great extent are dominated by the Western perspective. This is often underpinned by the assumption that CEE institutions need knowledge transfer and solutions from more developed North-West countries, or the imposition of specific goals of gender equality policies and aspects which they should cover.

The CoP facilitators successfully develop and uphold a stimulating and supportive space for communication and the exchange of ideas, knowledge and experience .... This space is especially valuable because of its unique focus on the CEE region, which rather often has been neglected in wider European discussion on gender equality in research and innovations. Thus, participation in the CoP's activities enriches us with knowledge about gender equality achievements in other RPOs in the region, expands our understanding of undergoing process, and reinforces our general sensitivity to the topic.

(Blog entry 2)

In the interviews and during the meetings, members reported receiving new knowledge, and perspectives on approaches, solutions and strategies on how to develop a GEP. In this context, the usefulness of workshops on undertaking concrete actions towards implementing GEPs and building argumentation for structural change was raised.

> ... we got some information about how to learn, how to analyse, how to design and implement a gender equality plan. So, I started to do that. For example checking my institutes' regulations and checking national regulation, and I started to do that with a handbook about practices (...), I think that I did concrete steps thanks to information, which I get during CoP meetings.
>
> *(Member 1, interview)*

The CoP members also appreciated being updated on gender equality resources and events and having access to different perspectives, as the CoP members represent various scientific disciplines and hold different positions in their organisations. Access to gender knowledge and expertise was seen as equally advantageous for the members themselves and for their organisations (also see Chapter 10):

> Yes and of course, this cooperation of a community. It gave me a lot as a person that I could meet these people and get some expertise. I have the feeling that when I need some help, or when I need some external expertise, I know where to go. This is very important for me, but also for [name of research centre].
>
> *(Member 2, interview)*

However, some limitations to acquiring gender knowledge has been observed as well. While transferring solutions and approaches within one type of organisation (i.e., from one university to another) was perceived easy and effective, knowledge transfer to non-university research institutions was seen as requiring adaptations and, therefore, more difficult.

Practical tools for collecting data, enhancing gender knowledge, and monitoring interventions

The analysis indicates that ready-made tools are beneficial for members of the CoP that often have limited resources for acting. A gender equality audit and monitoring (GEAM) tool for carrying out survey-based gender equality audits in academia and RPOs developed by the ACT consortium (for more details, see Chapter 3) proved most needed. It fitted well with the well-articulated need to conduct regular assessments of gender equality and organisational culture status quo. Therefore, most of the CoP members engaged in the preparation phase for the launching of the GEAM survey – they participated in the pilot survey and gave their feedback on its content and functionalities, they joined forces to translate the questionnaire and relevant documentation into local languages and adapt the questions to the specificity of their organisations and to the context created by the COVID-19 pandemic.

> Our motivation to join the project was to develop the discussion about gender equality, promote the idea of equality among university employees and encourage them to complete the survey 'The Gender Equality Audit and Monitoring Tool'. Our intention is to collect data on issues such as: stereotypes, prejudices, bias, organisational culture and climate, behaviour, interpersonal experiences etc. and share them with international academic environment. We also plan to deepen the analysis with a qualitative study on employees' personal experiences related to gender equality.
>
> *(Blog entry 4)*

While there were some concerns as to whether there would be the organisational will to use the tool and some delays due to the COVID-19 pandemic, as of June 2021, the survey has been already conducted in 12 of the member organisations. The GEAM was widely recognised as a tool for providing information on the situation in institutions and the basis for future evidence-based interventions, monitoring change, and improving working conditions in participating organisations. Running the survey and presenting its results were also reported to trigger a discussion on gender equality issues among employees and to help create allies and synergies within organisations. The comparability of results, both nationally, regionally, and internationally, was as well seen as the CoP's strength through which a CoP can potentially attract the attention of policymakers and foster a discussion about gender equality issues in research and academia in the CEE region.

However, not all members managed to obtain consent from their organisational leadership to conduct the GEAM survey. In some organisations, other surveys on gender equality and/or working conditions had been recently carried out or already scheduled. In others, the discussion on conducting the GEAM survey and implementing GE measures was put on hold due to the election of the university authorities, the COVID-19 pandemic, and institutional resistance towards gender auditing.

Apart from the possibility of using tools developed in the ACT project, members of the GEinCEE CoP co-created an online map of GE bodies and measures in their own institutions and gave accounts of the developments in relation to gender equality in the blog. The map presenting gender equality bodies and measures present in HE and R&I in the CEE region (https:// geincee.act-on-gender.eu/gender-equality-mapping) was an effect of coordinated action of assessing gender equality status quo at member organisations carried out in the first months of the CoP's operation. The map and the blog accounts were recognised as important and needed output from the collaborative activity within the CoP: it increased the visibility of gender equality issues in CEE and the CoP itself by providing data on the status quo of gender equality and addressing the gaps in the research concerning CEE countries. Additionally, the online map and the accounts are thought to be useful in advocating for progress in members' own organisations (by demonstrating developments in other places).

Dealing with resistance

The Community Mapping results and testimonies given by the GEinCEE CoP members proved that having effective tools to overcome resistance towards gender equality interventions is needed. While argumentation strategies for gender equality were more systematically tackled in the CoP's trainings, the community itself has been recognised as a safe space to discuss sensitive and difficult issues, including resistance to change. The CoP's meetings provided opportunities to both share good practices in dealing with resistance, already tested by some partners, and receiving emotional support by those who face reluctance or hostility towards gender equality interventions.

However, as the manifestation of resistance can take various forms, a CoP is not viewed as providing a solution to all problems. The context of anti-gender discourse and initiatives prominent in some of the CEE countries was perceived by the CoP members as having a negative impact on their work and the possibility to implement gender equality interventions. This inhospitable (or even hostile) climate in one of the member countries impeded the activities of a newly created intra-university gender equality committee. The team encountered hate which diminished their motivation for engagement and caused censorship in regard to language used to name their activities, in particular the reflection was made around the concept of "gender", raising heated discussions. In many institutions, the name of the survey was renamed using the concepts such as "equality between women and men", "equal treatment", or "work conditions".

> When the antidiscrimination team was launched last year, the information appeared on social media and local newspapers. The amount of hate the team faced was overwhelming and nobody expected that. It was not only criticism, or unnecessary statements. It was pure hate. It showed me that we can try to make this structural change and we can look for tools, but there are people simply not wanting to have this

change at all. They use methods, words and tools where we have no tools to fight with. How can you fight the hate? If you try to explain yourself, it brings even more haters into discussion. What happened with my colleagues in this team? They participated in launching the survey within the project, but whenever they have to place their names somewhere officially, they have to reconsider, if they really want to do it. We had a huge discussion about if we should use the word gender in the title of the survey. It will bring hateful emotions again. I think the limits of the CoP are in the reality and political atmosphere, which is created by the government.

(Member 2, interview)

First accounts of the CoP members, collected after the European Commission announced introducing a GEP as eligibility criterion in the application for the Horizon Europe programme[3], suggest that this decision should trigger changes, also in those partner organisations where organisational resistance had previously prevented any initiatives. However, the doubts are raised whether the requirement to have a GEP would help to improve organisational cultures of academia and research institutes, or its effects would rather be limited to signing a paper and ticking boxes on a checklist.

Framing gender equality within wider concepts and initiatives

Within the CoP, different strategies were discussed and applied relating to how to prioritise gender equality and manage resistance at member organisations. The CoP created a space to discuss the framing of gender equality in connection to other institutional strategies or plans (Palmén et al. 2019). The HR Excellence in Research Award was widely recognised among the CoP members as affecting the commitment towards advancing gender equality[4]. Some CoP members admitted that being a member of a CoP is another step to show the commitment to gender equality captured in the strategy on HR Excellence in Research, either by reinforcing the promised actions or initiating the discussion on gender equality on the institutional agenda. In this context, promoting the GEAM survey – as well as any gender equality interventions – as relating to the HR Excellence in Research action plan was seen as facilitating the involvement of organisational management to those activities. Additionally, as some members indicated, efforts related to assuring good work conditions and initiatives in relation to diversity and social responsibility, together with actions towards assuring gender equality, all exemplify a long-lasting commitment to capacity building and the incorporation of the voices of underrepresented groups, especially women.

As one of the first universities in Poland, we have implemented the Diversity Charter and the Declaration of Social Responsibility. We were also recognised with the HR Excellence in Research Award and have experts who deal with the social responsibility issues at the university. The UL [University of Łódź] participation in the GEinCEE CoP is a result of previous steps taken to manage our social responsibility and gender equality.

(Blog entry 6)

However, even with many positive voices towards linking gender equality to other strategies, there was no consensus on such strategies. There have been doubts voiced by CoP members about embracing gender equality arguments and interventions within other concepts, such as diversity or anti-discrimination. One objection referred to the risk of blurring or melting the idea and priorities of gender equality. The other doubt referred to national and European legal and policy frameworks, where anti-discrimination and gender equality are treated as separate concepts, with the second one requiring adopting an intersectional perspective.

Human and financial resources

Structural change in research institutions demands human and financial resources. In some of the national contexts, EU-funded structural change projects have initiated effective institutional change and have provided one of the main or the sole encouragements to setting up a GEP in research and HEIs (EIGE 2016). This is also reflected in the experiences of GEinCEE CoP members – those of them who have already participated in the FP7 or H2020 projects gained financial, time, and human resources to advance gender equality frameworks. A share of our members were in fact engaged in such projects and the GEinCEE CoP has become a space to discuss their experiences, such as in the case of University of Gdańsk (STARBIOS 2) where gender measures were implemented under the framework of Responsible Research and Innovation (RRI), or projects which focused on implementing specific gender equality measures (e.g., WLB policy implemented under LIBRA project at the Central European Institute of Technology – Masaryk University) or a GEP (e.g., at the Vilnius University under SPEAR project).

At the same time, the impact of the EU funding is problematic due to sustainability-related impact. In the GEinCEE CoP, it was the H2020 ACT project which provided financial and human resources for its operation through covering the costs of group facilitation, meetings, trainings, and materials. However, it did not include securing financial and human resources necessary to undertake gender equality initiatives at member organisations. In this context, a few members – especially those who were not formally engaged in other European structural change projects, reported the lack of dedicated funding as great limitation to their gender equality activities. While the CoP support programme enabled the provision of some service for willing organisations, a need for applying for external funding was articulated on various occasions, including the discussions on the sustainability of the CoP and financial resources for personnel involved in the development and implementation of gender equality measures.[5]

The embeddedness of CoP in the ACT project was seen as producing ambivalent results. The CoP members –beside the CoP facilitators – have not been formally engaged in the Horizon project which impacted on their identification and engagement with the project, as they *"aren't regular project members, actually, [they] are like guests" (Member 1, interview)*. On one hand, it leaves more room for spontaneity and enables the CoP to be inclusive for new members. On the other hand, it gives the impression that there are no binding rules, which might mean that some members are more inclined to skip meetings and avoid mutual engagement due to voluntary commitment.

Conclusions

The GEinCEE CoP aims at addressing the conditions for structural change in HE and R&I in CEE that resonate with the needs formulated by the CoP members. As we have shown, the CoP to some extent fulfils the role of an intermediary support structure, combining isolated efforts to implement change in the region. However, it cannot be seen as a remedy for all challenges encountered by gender practitioners and scholars. The analysis shows that the CoP provides possibilities for an exchange of knowledge and building up gender expertise, which has been evaluated as an effective way of supporting gender equality initiatives in member organisations that lack employees with such expertise. In the context of scattered activities, it allowed for effective experience and practice exchange, capacity building, and it took steps to gather the experience and knowledge from the region. That not only was important from the knowledge-management perspective but also to give voice to practitioners and experts from the region that may share different experiences to their Western counterparts. In the context of resistances and the minimal strategic placing of gender equality, it provided local experts with some extra-institutional embedding of their activities, as well as strengthened their personal and collective agency. But even if the CoP is recognised as a safe space for sharing difficult situations, its effectiveness in managing individual cases of resistance proved limited.

Similarly, the need for financial resources necessary for implementing gender equality interventions has been addressed to limited extent. The CoP financed face-to-face meetings and its support programme sustained selected activities, but it did not provide resources for e.g., hiring personnel for implementing gender equality interventions, which in many organisations is missing. However, the CoP became a platform for – previously lacking – regional and country-based cooperation allowing for forming project teams that can apply for external funding.

The challenge of lacking resources was to some extent addressed through an access to ready-made tools. The GEAM survey tool proved to be an important instrument facilitating gender equality initiatives by equipping their advocates with evidence-based legitimisation for the need of these interventions and by enabling regular monitoring of their effects. In some of the member organisations, the CoP helped to initiate concrete activities, or strategic reflections, but by itself it did not trigger substantial changes, rather supported ongoing initiatives. At this point we can see it more as a support mechanism to wider initiatives, that are triggered by other external forces, such as EU structural change projects or potentially in the future by the requirement to have a GEP to be eligible for participation in the Horizon Europe programme.

As seen by its members, through organised conferences, planned publications, and the dissemination of its results, the GEinCEE CoP has the growing potential to enhance the visibility of gender equality issues in CEE and reinforce a discussion on the specificity of the region in the European context. It has also become a platform for developing an agenda and methods that would include local settings and transcend a simple transfer of solutions developed in the Western European context. Trying to bridge scattered activities and connect regional actors however is not enough. Effective strategies to enhance its visibility to lobby national and regional policy makers for implementing legal and policy incentives for introducing gender equality measures are still needed.

Notes

1. Analysed blog entries are published at: https://geincee.act-on-gender.eu/activities.
2. For example, in the #COMMIT2GENDER campaign for #8M2020 that was coordinated by the ACT project and co-organised by members of 12 H2020 structural change projects to share their goals and commitment on gender equality.
3. The Participants of the Horizon Europe programme that are public bodies, RPOs, or HEIs established in an EU Member State or Associated Country must have a gender equality plan in place (applicable from 2022 onwards).
4. The HR Excellence in Research Logo is awarded to HEIs and ROs which have implemented the "HR Strategy for Researchers" based on adherence to the principles of The European Charter for Researchers and the Code of Conduct for the Recruitment of Researchers adopted by the European Commission. This strategy includes implementing equal opportunity policies at recruitment and promotion levels to obtain a representative gender balance at all levels of staff, including at supervisory and managerial level (European Commission 2005). As of June 2021, eight CoP member organisations had received the HR Excellence in Research award and two organisations were officially planning to apply for it.
5. See Chapter 10 of this volume by Reidl et al., who show that this is also an issue for the other ACT CoPs.

References

Böök, Birte, Susanne Burri, Linda Senden and Alexandra Timmer. 2021. *A comparative analysis of gender equality law in Europe 2020*. Luxembourg: Publications Office of the European Union.

Cavaghan, Rosalind. 2017. Bridging Rhetoric and Practice: New Perspectives on Barriers to Gendered Change. *Journal of Women, Politics & Policy*, 38(1), 42–63.

Colizzi, Vittorio, et al. 2019. Structural Transformation to Attain Responsible BIOSciences (STARBIOS2): Protocol for a Horizon 2020 Funded European Multicenter Project to Promote Responsible Research and Innovation. *JMIR Research Protocols*, 8(3), e11745.

EIGE. 2016. Integrating gender equality into academia and research organisations. Analytical paper, https://eige.europa.eu/publications/integrating-gender-equality-academia-and-research-organisations-analytical-paper, 10.06.2021.

European Commission. 2019. *She figures 2018*. Luxembourg: Publications Office of the European Union.

European Commission. 2005. *The European Charter for Researchers The Code of Conduct for the Recruitment of Researchers*. Luxembourg: Office for Official Publications of the European Communities.

Flyvbjerg, Bent. 2011. Case study. In: N. K. Denzin and Y. S. Lincoln (eds.). *The sage hand-book of qualitative research*, 4th ed. Thousand Oaks, CA: Sage, 301–316.

Hearn, Simon, and Nancy White. 2009. *Communities of practice: Linking knowledge, policy and practice*. London (UK): Overseas Development Institute.

Krizsán, Andrea, and Conny Roggeband. 2019. *Gendering democratic backsliding in Central and Eastern Europe. A comparative agenda*. Budapest: Central European University.

Lansu, M. Bleijenbergh and Yvonne Benschop Y. 2019. Seeing the System: Systemic Gender Knowledge to Support Transformational Change towards Gender Equality in Science. *Gender Work & Organisation*, 26(11), 1589–1605.

Linková, Marcela, Dunja Mladenić, Eszter Papp, and Katerina Saldova. 2007. *Gender issues in science as a luxury. Enwise follow-up activities in Central Europe, Central European Centre for Women and Youth in Science,* http://sciencewithart.ijs.si/pdf/GenderIssues_CEC-WYSComparativeReport.pdf, 28.04.2021.

Lipinsky, Anke. 2014. *Gender equality policies in public research. based on a survey among members of the Helsinki group on gender in research and innovation 2013*, Luxembourg: European Commission, https://op.europa.eu/en/publication-detail/-/publication/39136151-cb1f-417c-89fb-a9a5f3b95e87, 28.04.2021

Mühlenbruch, Brigitte and Maren A. Jochimsen. 2013. Only wholesale reform will bring equality. *Nature*, 495, 40–42.

Palmén, Rachel, Maria Caprile, Rosa Panadès, Julia Riesco, Elizabeth Pollitzer, and Claartje Vikenburg. 2019. Conceptual Framework – Gender Equality and Communities of Practice (Version 1). *Zenodo*. http://doi.org/10.5281/zenodo.3235296.

Palmén, Rachel, and Evanthia Kalpazidou Schmidt. 2019. Analysing Facilitating and Hindering Factors for Implementing Gender Equality Interventions in R&I: Structures and Processes. *Evaluation and Program Planning*, 77, 101726.

Pelenc, Jérôme, Didier Bazile, and Cristian Ceruti. 2015. Collective Capability and Collective Agency for Sustainability: A Case Study. *Ecological Economics*, 118, 226–239.

Reidl, Sybille, Ewa Krzaklewska, Lisa Schön, and Marta Warat. 2019. ACT Community Mapping Report: Cooperation, Barriers and Progress in Advancing Gender Equality in Research Organisations. *Zenodo*. http://doi.org/10.5281/zenodo.3247433, 28.04.2022.

Starman, Adrijana Biba. 2013. The Case Study as a Type of Qualitative Research. *Journal of Contemporary Educational Studies*, 1, 28–43.

Timmers, Tanya M., Tineke M. Willemsen, and Kea G. Tijdens. 2010. Gender Diversity Policies in Universities: A Multi-Perspective Framework of Policy Measures. *Higher Education*, 59(6), 719–735.

Verge, Tània, Mariona Ferrer-Fons, and M. José González. 2018. Resistance to Mainstreaming Gender into the Higher Education Curriculum. *European Journal of Women's Studies*, 25(1), 86–101.

Vinkenburg, Claartje J. 2017. Engaging Gatekeepers, Optimizing Decision Making, and Mitigating Bias: Design Specifications for Systemic Diversity Interventions, *The Journal of Applied Behavioral Science*, 1–23.

Warat, Marta, Ewa Krzaklewska, Paulina Sekuła, Sybille Reidl, and Lisa Schön. 2019. Implementing gender equality in academia in Central and Eastern Europe: Achievements, barriers and a way forward? A presentation for the conference Gender Studies and Research 2019: Centenary achievements and perspectives, Vilnius, November 21–23, 2019.

Yin, Robert K. 2014. *Case study research: Design and methods*. Thousand Oaks, CA: Sage.

6 Alternative infrastructure for gender equality in academic institutions in Slovenia

Community of Practice approach

Jovana Mihajović Trbovc

Introduction

Dealing with the problem of gender equality in the sector of higher education and scientific research has a rather long history in Slovenia, going back to the period when it was part of socialist Yugoslavia, which dissolved in 1991. The conceptual framework and vocabulary through which 'the problem' (Bacchi, 1999) of women in science and academia was described changed over time, as did political and socio-economic setting. Nevertheless, there is undeniable continuity in the efforts on these issues, demonstrated by long-term commitment of prominent individuals and institutional actors, as will be described in this chapter. Looking at the span of more than 30 years, there is a tendency of new initiatives and projects dedicated to improving gender equality in research and higher education in Slovenia to build upon achievements of prior initiatives and knowledge accumulated therein.

When the project team of the Research Centre of the Slovenian Academy of Sciences and Arts (*Znanstvenoraziskovalni center Slovenske akademije znanosti in umetnosti* – ZRC SAZU) was tasked with creating a Community of Practice (CoP) as part of the ACT project, it was most meaningful for us to enhance further this local exchange that proved fruitful over the years. Therefore, the CoP named Alternative Infrastructure for Gender Equality in Academic Institutions (Alt+G), which has been coordinated by the author of this text, should be observed in this longer history of efforts to achieve gender equality in Slovene academia. In this paper, I will demonstrate that the focus of transformative efforts shifted from the level of national regulations to the academic institutions, due to systemic conditions. Furthermore, I will argue that the CoP approach is particularly beneficial for spreading and multiplying structural change *within* the institutions of higher education and research organisations, and that it can help overcome certain systemic fallacies.

This chapter will first provide a historical overview of activities and actors that have been involved in the issues relating to gender equality in

DOI: 10.4324/9781003225546-6

Slovenian academia – the legacy on which the CoP Alt+G was built. Then, it will showcase how the focus of the group shifted from fixing national regulations towards institutional change. Finally, the paper lays out the benefits of the CoP approach for structural change within academic institutions.

An overview: from 'women in science' to 'gender equality' and beyond

While Yugoslav socialist system was based on egalitarian principles in different walks of life (economy, politics, education, etc.) it also created institutional foundations that enabled synchronisation of work and care (e.g., paid parental leave, affordable public nurseries, and kindergartens). Both these factors led to relatively large participation of women in workforce, including research and higher education,[1] though the numbers significantly varied across the country. Being a socialist country, outside of the Soviet bloc and a leader of the Non-Aligned Movement, Yugoslav official politics incorporated 'women's question' into the concept of 'social development' (Božinović, 1996), including the education policies (Tomšič, 1980: 149–155). However, since 1970s, nascent feminist movement in Yugoslavia started voicing criticism of the official state policies and political practices for their avoidance to address gender inequalities that were still persistent in the society (Lóránd, 2018). Yugoslav neo-feminists, as they were called, started pointing to discrimination of women in public and professional life, which had a reflection in some sociological studies of the time (e.g., Hvala, 1979; Jogan et al., 1986). Initially, Yugoslav sociologists dealt predominantly with inequalities (class, urban/rural and gender) in access to education, while the pioneer research into the position of women in academic professions appeared in Croatia (Prpić, 1982, 1989) and Serbia in the 1980s (Blagojević, 1990).

In Slovenia, in particular, the issue of the position of women in science, their underrepresentation at the top of the academic ladder as well as the sexism and discrimination they are subjected to, was raised both as a scholarly and public policy problem around the time the country became an independent state in 1991. At that time, first pilot studies on women in the Slovene academic field emerged (Jogan, 1992) and the public discussion started among professionals: the National Commission of UNESCO organised a roundtable on women in science and research during the series of public events called Science Week in autumn 1994 (Umek Venturini, 2015, p. 19). This body also financed the first whole-scale research on the population of female academic staff in Slovenia in 1996 (ibid; Jogan, 1998), which was led by Maca Jogan, professor of sociology at the Faculty of Social Sciences, University of Ljubljana. She was the driving force of many initiatives targeting the improvement of the position of female scholars, even after ending her active academic career, and remained avidly active as speaker or onlooker on many public events to this day.

The growing importance of this topic could be illustrated by the attention it was given within the local professional community. In 1997, the magazine issued by the then Ministry of Science and Technology called *Raziskovalec* [Researcher, in male form] dedicated a special issue to the topic of women in science (vol. 27, no. 1–2). Some of the individuals that will remain active on the topic wrote opinion-pieces on this issue, while the contributor to the editorial, Tamara Lah Turnšek, biologist and then director of the Institute of Biology, was named by the Ministry as the first National Coordinator for Women in Science in 1999. She was also a delegate to the Helsinki Group on Women in Science that was established in 1999 – a special group of 15 EU member and 15 non-member states whose 'aims were to stimulate research on women's position in science, to coordinate the debate about the policy and measures regarding the promotion of women in science, and to search for statistically sensitive gender indicators' (Jogan, 2006, p. 46). In sync with activities of the Helsinki Group, the Slovene Ministry founded a consultative body first named the Commission for the Recognition of the Role of Women in Science [*Komisija za uveljavitev vloge žensk v znanosti*] in 2001, which was a decade later renamed as the Commission for Women in Science and as of 2018, Commission for Equal Opportunities in Science.[2] The changes in the naming of the Commission reflect the larger shifts in the perception and conceptualisation of the problems related to the issues of equality, equity and gender in academic institutions in Slovenia: shifting from binary to a more fluid plural notion of gender, and adopting concern for equal opportunities of various marginal groups (young, foreigners, etc.).

The Commission consecutively gathered experts dealing with the issues related to gender equality in academia, as well as prominent scholars from different disciplinary fields dedicated to advocacy for (more) women in science, all of whom have been engaged on a voluntary basis. While the Commission is usually asked to comment drafts of key policy-making documents (e.g., 5-year National Programme for Equal Opportunities between Women and Men; 10-year Research and Innovation Strategy); it does not have executive nor legislative power of any kind. This practically means that the Commission has rather symbolic power, and is not apt for driving systemic change on its own. Though it is promoted by the government as a flagship of its endeavours in the field of gender equality in academia, it actually functions rather as a group of activists and engaged researchers. I argue that the very existence of the Commission for Women/Equal Opportunities in Science is often used as an excuse for the lack of systemic support and action on the part of the government.

An illustration for this argument is the fate of the most comprehensive measure that was envisioned by the policymakers but was never executed: the *Resolution on National Research and Development Strategy of Slovenia 2011–2020* sets out for Ministry of Education a task to create an 'Action Plan for Improving Career Opportunities for Researchers in all Career Periods and for Ensuring the Gender Equality Principle' (ReRIS11-20, Chapter 4.1, goal no. 5, measure no. 34) – to be executed until 2012. The creation of the Action Plan was postponed until 2017 and, later abandoned, while other policy documents were adopted, seemingly complementing the commitment set out in the Strategy. However, the essence of the policy initiative has changed in the process: Slovenian ERA Roadmap 2016–2020 mentions the Action Plan (with the same title) not as a document to be prepared (and executed) by the Ministry of Education, but rather as a document that academic institutions should create by and for themselves – in other words, a gender equality plan (goal no. 18, p. 19). In this way, not only terminological (and policy) confusion has been created, but also the Ministry avoided to take on its share of responsibility. While the Slovenian ERA Roadmap uses as an indicator of success a number of research performing organisations obtaining a GEP, it does not set out any concrete support to this end – which should have been the essence of the above mentioned Action Plan (Mihajlović Trbovc, 2021a).

In early years, the Commission was mostly dedicated to raising awareness on the issues of gender (in)equality in academia, while some of its prominent members and presidents conducted research on their own (e.g., Maca Jogan, Mirjana Ule, Milica Antić Gaber). The Commission influenced changes in some of the relevant regulations, such as the Rules on the Procedures of (Co)financing, Evaluation and Monitoring of Research Activities, a national bylaw which sets the standards for evaluating scientific excellence that are applied both by research performing and research funding organisations. The Commission instigated the change in the Rules pertaining to the procedures for the evaluation of researchers, meaning that the five-year period for assessing scientific excellence is extended for the time period researchers spent on leaves (such as one-year parental leave) – thus, ensuring more gender equity in the evaluation of project applicants. Similarly, the time limit for applying for PhD positions and postdoctoral projects is extended one year for candidates who used their parental leave or other leave of absence. In this way, the Commission expanded to the national level an endeavour started by Maca Jogan, its long-term co-president (2001–2009), who initiated such change in regulations within her own institution – Faculty of Social Sciences, University of Ljubljana – already in the 1970s (Jogan, 2006, p. 29). Another sphere of early activity of the Commission for Women/Equal Opportunities in Science has been promoting gender-sensitive research. This endeavour however has not been sustainable and proved limited due to lack of resources (on the part of the body) to conduct or commission systemic and regular research on its own.

Over the years, its annual and occasional events became a stage for Slovenian members of European projects consortiums to disseminate results and their project activities (e.g., Genis Lab, GARCIA, PLOTINA, etc.).[3] This is how the information about concrete gender equality measures that could be applied within research organisations has been spread within Slovenian academic community. However, what turned out to be a recurring phenomenon is that in most cases these project activities ended with the lifespan of a European project, and usually no sustainable institutional change has been achieved. Furthermore, up until 2022, no institution in Slovenia had established a gender equality office or a gender equality officer (as a paid position) that would be in charge of constant supervision of these issues within an institution.

For this reason, the ZRC SAZU team within the ACT project embarked on making use of this accumulated knowledge and experience by connecting individuals who have engaged with the activities of the Commission for Women/Equal Opportunities in Science and/or had experience with European projects dealing with gender equality in academia. The idea was to build a community of scholars and administrative staff from different academic institutions in Slovenia who are interested in implementing gender equality measures at their institutions. The CoP was named Alternative Infrastructure for Gender Equality in Academic Institutions (Alt+G), since its main focus is on making an alternative institutional infrastructure for sharing knowledge, experiences and strategies for implementing gender equality measures at both institutional and national level.

Before the formal establishment of the Alt+G CoP in summer of 2019, the CoP was in the 'inquire' lifecycle phase (Cambridge et al., 2005)[4] although some preliminary meetings between ZRC SAZU team and potential CoP members already took place. In that sense, it was a new community yet based on the core of a certain number of individuals and institutions which already have cooperated. The community Alt+G gathers 37 individuals from 16 institutions.

From fixing national regulations towards institutional change

The systemic conditions in the Slovenian academic field influenced the focus of the initial activities of the Alt+G CoP members. For instance, the standards for evaluating academic excellence and norms for career promotion are regulated by national laws and bylaws, which are then applied (and translated into concrete practice) by the individual research organisations. Therefore, the CoP members initiated several activities that aimed at changing regulations in such a way that the standards are easier to be achieved by researchers who are burdened by caring duties at home, which are in most cases women having young family in the middle of their career progress.

The first initiative was intended to intervene on the Rules on Academic Research Titles, a national regulation applied by the research organisations for evaluating advancement through the ladder of academic titles for individual researchers. The Alt+G CoP members pointed to the problem that many researchers, particularly women, were struggling with fulfilling the key requirement for advancing from junior to senior research position – that is concluding mentorship to a PhD student. This proved to be a particular challenge to researchers working at research institutes, which have a scarcity of PhD positions, as well as teachers at universities, due to the informal practice (widespread in Slovenia) that heads of departments (or research groups)[5] have a say on who will get a mentorship – systemic problems that were already discussed among professionals (Adam, 2015) and in the media (Škerl Kramberger, 2017). Furthermore, the institutions as a rule do not have systemic measures in place to financially (or otherwise) support *all* of its researchers in attaining a mentorship in time for the regular five-year evaluation. Therefore, researchers are usually left to their own devices to navigate intra-group dynamics (e.g., lobbying the head of the department or the head of their research group) or fundraise for a paid PhD position.

Though this is not a gendered issue per se, the lived experience of CoP members testifies that the unwritten rules of decision-making within research groups and university departments often sidestep women or relegate them to less attractive jobs that put them further away from fulfilling this particular requirement. For instance, several CoP members reported knowing a case that a head of a research group (by the rule a senior experienced researcher, more often male than female) arbitrarily decided to take for himself the one financed mentorship the group obtained through a public call, rather than allowing younger group members to have a chance to be a PhD mentor for the first time in their careers. The claim that such informal practice exists was confirmed also outside Alt+G group during the workshop with stakeholders organised by CoP as elaborated further below. Even the official analysis of annual data confirms such informally gathered information: in year 2020, 25.5% of all publicly financed mentorship positions (that is 256 out of 1004) in Slovenia were distributed to the heads of the research groups (ARRS, 2021).

The lived experience of CoP members is also confirmed by previous studies that mapped out overt and obscure discriminations (Ule, 2013) and gendered stereotypes women face in the Slovenian academy, as well as their overburdening at home and work (Jogan, 1998, 2006; Ule 2012). Previous studies also pointed to implicit (Luthar & Šadl, 2002) and structural (Antić Gaber, 2010) ways in which institutions and the whole academic system reproduce gendered hierarchies (Ule et al., 2013).

Therefore, the Alt+G CoP members first discussed the problem within the group, sharing real-life examples from their own and their colleagues' experiences. After deliberation, the CoP members detected that an important part of the problem lays in the problematic practice of the Scientific Council of the Slovenian Research Agency (*Agencija Republike Slovenije za raziskovanje*, ARRS), which has the power to initiate amendments to the Rules on Academic Research Titles and apply them as an instance of the appeal.[6] Then, the Alt+G members decided to act by issuing a petition to the Scientific Council and created a smaller working group within the CoP for this purpose, which was active throughout 2020. The group requested relevant information from the ARRS administration (e.g., on number of appeal cases) and a group member made an analysis (Opara Krašovec, 2020) and drafted the petition. Then they gained the support of the Commission on Equal Opportunities in Science, and obtained a meeting with the Scientific Council members, which took place on 14th September 2020. The petition contained suggestions for long-term and short-term interventions. The long-term goal was to change the Rules on Academic Research Titles in such a way that the requirement of (successful, that is finalised) PhD mentorship could be relaxed and/or substituted with other similar criteria. The second short-term aim was to influence the practice of Scientific Council deliberation in appellate cases is more flexible within the existing requirements. Although the initiative was not successful in changing the wording of the regulation, it managed to get assurance of the Scientific Council to adopt new principles in evaluating this requirement.

The engagement in this joint activity encouraged CoP members to consider what kind of changes are needed at the level of academic institutions, so they could enable *all* of their researchers to fulfil criteria for academic advancement in a relatively equal manner, and not be indirectly discriminated by the rules of the game due to their position within the academic hierarchy or their gender. This was the topic of the workshop titled 'Criteria of academic career progress in Slovenia' that gathered CoP members and different stakeholders, such as trade union representatives and experts from the national accreditation body.[7] The workshop resulted in the list of mapped problems, which are creating unfair inequality in the pace of academic advancement among researchers, and the list of recommendations for needed changes on a systemic and organisational level.[8] The big part of the discussion during the workshop was dealing with untangling these two levels of each complex problem: which part of it could be solved by changing national regulation, system of financing or official bodies, and which part of the problem should be solved by changing regulations and practices of individual academic institutions.

The intertwining of national regulation and everyday practices in academic institution appeared as an issue in yet another problem detected by the members of the CoP Alt+G: use of gender-sensitive language in the official communication of the academic institutions. In the exploratory survey that the facilitator organised at the beginning of the CoP's existence (October 2019) in order to assess what are the interests of group members, 11 out of 24 respondents expressed interest in engaging with this issue. Since the Slovenian language possesses elaborate morphological means to express grammatical gender, the language policy in this area is relevant. What frustrated many of the CoP members was the situation that their institutions (still) had a tendency to use the masculine form instead of feminine ones when referring to women's academic titles or a job position – in formal documents or public websites. To illuminate the problem, one should note that use of feminine forms for job positions (and academic titles) for women (i.e. social femininatives) became a standard in the Slovenian language as early as 1960s (Doleschal, 2015), and as of the 1990s, the use of the masculine form as a generic (and gender unmarked) is increasingly understood as sexist language (Kranjc & Ožbot, 2013).

In official documents, however, the masculine form is used as gender neutral, both in general documents (such as institutional regulations) and documents referring to particular individual (such as employment contracts) including the cases when the addressee is not male. Even more, some institutions are using only the male form in the online profiles of their employees, due to the technical solution of having a male-only blueprint for a profile page. This made women researchers feel 'invisible' and that being a female is 'somehow less worthy', as was stated during the first CoP meeting on this issue (3rd March 2020). Therefore, the Alt+G CoP members from seven institutions engaged in a working group dedicated to this issue. The first step was to collect information from HR offices and the administration of the institutions on why this practice is in place. The rather unison explanation was that the official documents need to use academic and job titles as prescribed in national trade union agreements for the sector of higher education and research, which all use only the male form. Therefore, it seemed that the only solution was to change the regulation at the national level.

However, a closer inspection of these national trade union agreements conducted by the CoP members revealed telling exceptions from this universal use of male forms: some job positions were given only in female form – and these were always professions marked with low financial, social and symbolic status, such as cleaning-*lady* [*čistilka*], washing-*lady* [*perica*], she-dactylographer [*strojepiska*]. In cases where there were male and female counterparts, the latter by the rule referred to a lower paid position, for instance: 'general secretary of university' [*generalni tajnik univerze*] referring to the administrative head is given in masculine form, while the position of 'secretary to the university leadership' [*tajnica vodstva univerze*] referring to an office assistant job is given only in feminine form. Furthermore, CoP members established that there are no legally binding reasons forbidding institutions to change the ongoing practice, rather that it is a matter of bureaucratic automatism. Therefore, some CoP members managed to incorporate into their gender equality plans (GEP) measures relating to gender-sensitive use of language in official documents, as in case of the Science and Research Centre of Koper (*Znanstveno-raziskovalno središče Koper* – ZRS Koper) which adopted its plan in early 2021.[9] Other CoP members focused on measures fostering use of gender inclusive language [*spolno vključujoč jezik*], such as use of expressions that are neutral or gender non-defining; use of specific graphic solutions (such as underscore) that suggest inclusiveness of all genders. Such measures are for instance incorporated in GEPs adopted by the Faculty of Arts, University of Ljubljana (*Filozofska fakulteta Univerze v Ljubljani* – FF UL) in May 2020, and the University of Maribor (*Univerza v Mariboru* – UM)[10] in October 2021.

For Alt+G CoP members, creation of a gender equality plan (GEP) was not part of the initial agenda. In the exploratory survey organised by the facilitator in order to assess CoP members individual interests (conducted in October 2019), 7 out of 24 respondents expressed interest but most of them were not ready to engage with this issue. There is a reason for this: participating institutions did not formally commit to create a GEP,[11] and the CoP members were in general not acting as 'official representatives' of their institutions. The large majority of CoP members (33 out of 37) are scholars who are not in decision-making positions at their institutions, and among those who are, only two are gender experts. Therefore, Alt+G members had personal rather than professional motivation to be part of a CoP – this gave them much freedom of speech within the group, but little leverage against inhouse resistances.

However, once the GEP has become a requirement in the application for the Horizon Europe programme as of calls opened in 2021 (and an eligibility criterion as of 2022), the situation significantly changed. Many of the CoP members' institutions initiated the process of creating a GEP, and they relied on CoP members as their key actors in the process.

CoP approach and structural change within academic institutions

While in the first year and a half of Alt+G existence, the focus of the CoP activities was on joint initiatives towards structural change, since the beginning of 2021, the exchange in relation to creating GEPs became the predominant part of the group members activities. Therefore, the CoP organised an online workshop on 18th February 2021 titled 'Gender Equality Plan for academic institutions: why and how', where CoP members shared their experience and information. It was open to a wider audience and attended by 60 individuals from different Slovenian institutions. The facilitator presented basic information about the necessary elements a GEP should contain in a blog written in Slovene that was widely circulated (Mihajlović Trbovc, 2021b). In addition, there was series of workshops about GEAM survey (see Chapter 3) organised by ACT project, GE Academy and Alt+G, from April to June 2021. The communication within the group mainly took place via mailing list, with occasional meetings between the facilitator and individual group members on zoom or live, when COVID-19 related restrictions allowed.

Before the onset of the rule that made GEP a requirement for participation in Horizon Europe in 2021, only two research performing organisations in Slovenia have implemented a gender equality plan – Research Centre of the Slovenian Academy of Sciences and Arts (ZRC SAZU) and the Faculty of Arts, University of Ljubljana (FF UL) – both of them are CoP Alt+G members, while the former is home to the CoP facilitator. Several CoP members were previously involved in European projects which resulted in creating some gender equality mechanisms at their home institution. For instance, National Institute of Biology tested the process of gender-blind hiring as part of CHANGE project (CHAlleNging Gender (In)Equality in science and research, www.change-h2020.eu) in 2019. As pointed out before, these were not always translated into sustainable long-lasting institutional change and were rather activities that ended within a lifespan of a project. As testified by a CoP member from the National Institute of Chemistry (and a member of the Commission for Equal Opportunity in Science), Marta Klanjšek Gunde, activities relating to gender budgeting were organised at her institution as part of the project Genis Lab (The Gender in Science and Technology, 2011–2014), while training and pilot project on gender dimension in research was conducted during PLOTINA project (Promoting Gender balance and Inclusion in Research, Innovation and Training, 2016–2020) – but these have not become regular and recurring activities within the organisation. Some of the CoP members initiated some institutional change before and separate to the gender-equality project. For instance, in May 2018, the FF UL changed the existing practice of writing its official regulations exclusively in the masculine form, and substitute it with the new rule in which feminine forms are also considered gender neutral/inclusive. The aim of the measure is to contribute to a more inclusive communication practice and break up with the existing norms that make non-male genders invisible. To conclude, these gender equality measures vary in their form and comprehensiveness, and cover different topics, not always overlapping with the areas to be covered by recommendations of the Commission (European Commission, 2021).

This was the starting point before Alt+G members got increasingly involved in creating concrete GEP documents and dealing with practicalities relating to feasibility of concrete measures. Here, CoP approach enabled several types of exchange.

First, there was a direct transfer of knowledge and experience between CoP members from different institutions. Sometimes, such an exchange would start by a CoP member posing a question on the Alt+G mailing list that some other group member would answer. More often a group member would directly ask the facilitator, who then made 'a match' with another group member that could help in a particular issue. This was beneficial for CoP members in finding a practical solution to a concrete problem and often resulted in 'mirroring' activities – that is, one institution copy-pastes measures from another institution. For instance, when looking for a proper form in which to publish its GEP, ZRS Koper to a large extent replicated the design created by the ZRC SAZU, while the content was different.

Another type of exchange took the form of unstructured sharing of experience on problems and challenges, during online discussions or via email threads on Alt+G list. Rather than seeking a solution to a concrete problem, this form of sharing was spontaneous, unpremeditated and often more personal. This was beneficial for reflection on the process of initiating and executing gender equality measures, such as how to avoid so called 'box-ticking' in fulfilling GEP Horizon Europe requirements, or how to take into consideration difference in the needs of administrative and research staff (Mihajlović Trbovc et al. 2022; Petrović, 2021). Even more, in this way CoP members gave each other 'moral support' when facing resistances within their institutions, such as conservative individuals within the team for creating GEP.

Finally, the third type of exchange was aimed at prototyping solutions to particular problems. Rather than exchanging ready-made solutions, this is a joint activity in which different group members add bits and pieces of their knowledge and reflections in iterative form of communication, so the problem solving takes place collectively. For instance, this is how reflections of CoP members influenced translation of the GEAM survey (version 2) into Slovenian language, which required considerable contextual adjustments for the specific academic setting and concern for gender-sensitive language. Through the form of workshops and consultation process the coordinator of the survey translation, Ana Hofman, a researcher from ZRC SAZU (and institutional coordinator for ACT project), collected opinions and reflections on the applicability of the survey in Slovenian context, including national welfare system, the specificities of working environment codes and practices in different types of academic institutions, concepts from gender studies and gender-sensitive expressions, etc. Actually, with every iteration the final output was slightly modified and brought closer to the demands of local academic environment. At the moment, different institutions in Slovenia, as well as CoP members, are considering execution of the GEAM survey, while part of it was conducted at the ZRC SAZU in September 2021.

The CoP as a malleable form of method for sharing knowledge seems to be particularly fit to make use of the contingent windows of opportunity for institutional change. The fixed structure and community identity of a CoP enables the group to get mobilised very fast, share ideas and transform them into action.

For Alt+G CoP first such occasion was a joint reaction regarding a promotional video made by the public agency SPIRIT Slovenia (Public Agency for Entrepreneurship, Internationalization, Foreign Investments and Technology), which group members found to be sexist: The video titled 'Slovenia. Green. Creative. Smart.' promoting local opportunities for foreign investment was filmed and directed in such a way that it presents only men as agents of development and innovation while women are portrayed exclusively in nonprofessional roles. Pointing to the problem of gender stereotypes, the Alt+G, as group of individuals, issued a public statement on 17 January 2020, which was then further covered by the media. The topic provoked quite a discussion in which CoP members expressed their opinions, heard the opposing opinions, and were ready to alter their starting position in face of a better argument. Some CoP members who did not have experience with social activism learned a few 'tricks of the trade' (e.g., how to make a statement critical enough and diplomatic at the same time). Some CoP members shared what they learned on public relations trainings (e.g., when is the best time of the day to send a public statement to the media). This was an occasion for sharing and deepening transgenerational and transdisciplinary transfer of both practical and scholarly knowledge on gender stereotypes – including 'feminist awareness' (in the meaning of being aware of gender inequities).

Another occasion when outside events made the Alt+G group interact in an unplanned way was a wave of metoo# movement in the academic sphere in Slovenia that started at the turn of 2021. Then in short time span, two instances of sexual harassment by University of Ljubljana professors were made public and eventually led to their suspension. The public discussion that ensued revealed a lack of regulation and procedures for sanctioning sexual harassment, especially when victims are students.

At the outset of the Alt+G existence, the issue of harassment was not part of initial discussions regarding the CoP domain, and was not included in the exploratory survey that the facilitator organised in October 2019, therefore no joint activity in this regard was planned. Two institutions did have this topic on their agenda: ZRC SAZU was committed to establishing an anonymous channel for reporting sexual harassment (in GEP adopted in February 2019), while the Faculty of Arts, University of Ljubljana, planned to develop mechanisms for dealing with cases of sexual harassment (in GEP adopted in May 2020). However, only after the cases of sexual harassment became public in December 2020 and January 2021, the other institutions and CoP members expressed interest in sharing knowledge and information on concrete practices related to this issue.

The individual CoP members from the Faculty of Arts, University of Ljubljana, being part of the group for implementing GEP at their institution, and learning from the experience of their institution responding to the first case of sexual harassment, participated in developing a protocol for reporting and sanctioning sexual harassment. This resulted in crafting the professional guidelines for prevention of sexual harassment, bullying and violence (Podreka et al. 2021) that established the system of trained 'confidential persons' [*zaupne osebe*] to whom students and employees could securely report a misconduct. The analysis of the university regulation and national legislation that was conducted (in part) by the CoP members from the Ljubljana Faculty of Arts, was valuable information for other CoP members when they started examining existing regulations and procedures (and lack thereof) at their home institutions. Furthermore, discussion within the group pointed to the different challenges pertinent to the university structure and the structure of the research institute. Therefore, the existence and structure of the Alt+G CoP enabled fast transfer of knowledge and practices within Slovenian academic community, which enhanced quality of institutional changes that took place in individual organisations.

Conclusion

Looking at the history of the endeavours for gender equality in the academic community in Slovenia, one could observe a continuity in sharing and passing on knowledge on the problems at stake and mechanisms for overcoming them. While the Commission for Women/Gender Equality in Science (existing since 2001) was a central actor in this process, its positionality and mandate (as advisory body to the government) did not make it fit for instigating and driving institutional change on the level of individual organisations. However, the CoP Alt+G built on the experience and knowledge accumulated within the Commission, while some individuals are members of both.

Due to the systemic conditions in which academic sphere of Slovenia operates, the Alt+G initially focused on altering national regulations and bodies in the sphere of science in order to enable fair career advancement for all academic staff. Instead of focusing on outright gender discrimination, the CoP members tried to change the unwritten rules of the game in local academic competition that often sidestep women. In the process, it became clearer how the national level regulations are intertwined with everyday practices in academic institution and their bureaucratic inertia.

The true value of the CoP approach became visible when the Alt+G members shifted from changing national regulations to institutional change in part intensified by the EU-wide policy making a GEP an eligibility criterion for participation in the Horizon Europe programme as of 2021. When group members increasingly got involved with envisioning concrete GEP measures within their institutions, the CoP approach enabled several types of exchange. (1) The direct transfer of knowledge and experience between pairs of group members resulted in benefits for finding practical solution to concrete problems. (2) The unstructured sharing of experience among CoP members on problems and challenges they faced in everyday work helped them reflect on the process and functioned as a 'moral' support when facing resistances. (3) Prototyping solutions to particular problems in iterative form of communication makes problem solving a collective endeavour improving the applicability of concrete practical measure.

On the occasion of unplanned outside events, the CoP structure and sense of community provided the framework that turned unforeseen challenges into windows of opportunity for institutional change, and created space for mutual learning. The existence and structure of the Alt+G CoP enabled fast transfer of knowledge and practices within Slovenian academic community, which enhanced the quality of institutional changes that took place in individual organisations. Since the CoP approach operates on the fuel of personal motivation and depends on individual rather than institutional commitment, its ability and reach in enhancing concrete institutional change is contingent on favourable structural context.

Notes

1. For instance, in 1976, women constituted 27,9% of all researchers employed in research performing organisations in Yugoslavia, while the share of women was significantly lower in the United States of America (6%) and Norway (11%) in comparative data (for year 1974) (Prpić 1982, pp. 55, 57).
2. The author of this text is a member of the present composition of the Commission for Equal Opportunities in Science.
3. Information on the mentioned projects: Genis Lab (The Gender in Science and Technology, 2011–2014); GARCIA (Gendering the Academy and Research: Combating Career Instability and Asymmetries, 2014–2017); PLOTINA (Promoting Gender Balance and Inclusion in Research, Innovation and Training, 2016–2020).
4. The 'inquire' lifecycle phase means that "through a process of exploration and inquiry, … the audience, purpose, goals, and vision for the community" are identified (Cambridge, Kaplan & Suter, 2005, p. 2).
5. By 'research group', here I am referring to the officially registered group of researchers who are working within one research programme [*raziskovalni program*] – a form of a long-term financing of joint collaboration, that is evaluated (and if successful extended) every five years.
6. When a candidate for advancement to a senior research title is not formally fulfilling all requirements prescribed by the Rules on Academic Research Titles, their employer may support their advancement, in which case the final decision lays with the ARRS Scientific Council.
7. The workshop was hosted by the CoP member Ana Rotter from National Institute of Biology, in cooperation with the project CHANGE, on 15th September 2020. Since the host was from Marine Biology Station, the workshop was organised in a neighbouring maritime town Izola/Isola, and due to restrictions related to COVID-19, the gathering took place on an open-air boat (Alt+G Blog, 2020).
8. Some of these recommendations are in line with the concrete GEP measures adopted by some CoP member institutions afterwards.
9. Another Alt+G institutional member and home institution of the facilitator, Research Centre of the Slovenian Academy of Science and Arts already had this measure in its GEP adopted in February 2019.
10. University of Maribor is registered as one legal body, while at University of Ljubljana each faculty is a separate legal body.
11. This issue was not covered in the Memorandum of Understanding with which academic institutions supported participation in the CoP.

References

Adam, Frane (2015). *Raziskovalna politika in dileme evalvacije v znanosti* [Research policy and the dilemmas of science evaluation]. Ljubljana: Inštitut za razvojne in strateške analize.

Alt+G Blog. (2020). A workshop on scientific promotion criteria – on a boat! Retrieved from https://altg.act-on-gender.eu/Blog/workshop-scientific-promotion-criteria-boat [20.10.2021].

Antić Gaber, Milica (2010). Akademske institucije v Sloveniji in (ne)enakost spolov [Academic institutions in Slovenia and gender (in)equality]. *Javnost – The Public, Journal of the European Institute for Communication and Culture*, 10(1), 75–83. doi: 10.1080/15358590903248751

ARRS. (2021). Pregledi in analize: Obseg in struktura financiranja: Mladi raziskovalci [Overviews and analyses: Volume and structure of the funding: Young researchers]. Retrieved from http://www.arrs.si/sl/analize/obseg01/mr.asp [20.10.2021].

Bacchi, Carol Lee (1999). *Women, policy, and politics: The construction of policy problems.* London: Sage.

Blagojević, Marina (1990). Društveni položaj žena stručnjaka u Jugoslaviji [Social position of women professionals in Yugoslavia]. Doctoral thesis, University of Belgrade.

Božinović, Neda (1996). *Žensko pitanje u Srbiji u XIX i XX veku* [Women's question in Serbia in 19th and 20th century]. Beograd: Devedesetčetvrta; Žene u crnom.

Cambridge, Darren, Kaplan, Soren, & Suter, Vicki (2005). *Community of Practice design guide: A step-by-step guide for designing & cultivating communities of practice in higher education.* EDUCAUSE Learning Initiative (ELI) & Virtual Learning Environment (VLE). Retrieved from https://library.educause.edu/resources/2005/1/community-of-practice-design-guide-a-stepbystep-guide-for-designing-cultivating-communities-of-practice-in-higher-education [20.10.2021].

Doleschal, Ursula (2015). Gender in Slovenian. In Marlise Hellinger & Heiko Motschenbacher (Eds.), *Gender across Languages: Volume 4* (pp. 335–68). Amsterdam: John Benjamins Publishing Company.

European Commission. Directorate General for Research and Innovation. (2021). *Horizon Europe Guidance on Gender Equality Plans.* Luxembourg: Publications Office. Retrieved from https://data.europa.eu/doi/10.2777/876509 [22.6.2022].

Hvala, Ivan (Ed.). (1979). *Društveni položaj žene i razvoj porodice u socialističkom samoupravnom društvu* [Social position of a woman and development of the family in socialist self-managing society] *(Portorož, 18–20.3.1976).* Ljubljana: Komunist.

Jogan, Maca (1992). Career opportunities for women scientists and visible and invisible sexism in Slovene society. *Higher Education in Europe,* 17(2), 107–23. doi: 10.1080/0379772920170216

Jogan, Maca (1998). Akademske kariere in spolna (ne)enakost [Academic careers and gender (in)equality]. *Teorija in praksa,* 35(6), 989–1014. Retrieved from https://www.dlib.si/details/URN:NBN:SI:doc-H4DHUIIL [20.10.2021].

Jogan, Maca (2006). My life as a (female) Sociologist. In Mike Forrest Keen & Janusz Mucha (Eds.), *Autobiographies of transformation: lives in Central and Eastern Europe.* Abingdon: Routledge.

Jogan, Maca, Fischer, Jasna, Končar, Polonca, Bučar, Maja, Rener, Tanja, Boh, Katja, & Milošević-Arnold, Vida (1986). *Ženske in diskriminacija* [Women and Discrimination]. Ljubljana: Delavska enotnost.

Kranjc, Simona, & Ožbot, Martina (2013). Vloga spolno občutljivega jezika v Slovenščini, Italijanščini in Angleščini [Role of the gender-sensitive language in Slovenian, Italian and English]. In Andreja Žele (Ed.), *Družbena funkcijskost jezika (vidiki, merila, opredelitve)* [Social functionality of language (aspects, norms, definitions)] (pp. 233–9). Ljubljana: Znanstvena založba Filozofske fakultete.

Lóránd, Zsófia (2018). *The feminist challenge to the socialist state in Yugoslavia.* Cham: Springer International Publishing.

Luthar, Breda, & Šadl, Zdenka (2002). Skriti transkripti moči: dominacija in emocije v akademski instituciji [Hidden transcripts of power: Domination and emotions in academic institution]. *Teorija in praksa,* 39(1), 170–95. Retrieved from https://www.dlib.si/details/URN:NBN:SI:doc-9KF78615 [20.10.2021].

Mihajlović Trbovc, Jovana (2021a). Kdo je zavezan narediti načrt za enakost spolov? [Who is obliged to create a gender equality plan?]. Retrieved from https:// spolinznanost.zrc-sazu.si/kdo-je-zavezan-narediti-nacrt-za-enakost-spolov/ [20.10.2021].

Mihajlović Trbovc, Jovana (2021b). Načrt za enakost spolov kot pogoj za projekte Obzorje Evropa [Gender equality plan as a requirement for Horizon Europe projects]. Retrieved from https://spolinznanost.zrc-sazu.si/novice-in-dejavnosti/ [20.10.2021].

Mihajlović Trbovc, Jovana, Černič Istenič, Majda, Petrović, Tanja, & Andreou, Andreas (2022). Structural Positions, Hierarchies, and Perceptions of Gender Equality: Insights from a Slovenian Research Organisation. *Družboslovne razprave*, 38(99), 103–28.

Opara Krašovec, Urša (2020). Zaključeno mentorstvo doktorandom kot pogoj za napredovanje na JRZ [Completed mentorship for PhD students as a requirement for career advancement at RPOs] (a working document), 25 November 2020, via mailing list Alt+G.

Petrović, Tanja (2021). Improving female researchers' careers through gender equality plan actions: Experiences from a Slovenian research institution. *Public Policy and Administration*, 20(1), 45–57. doi: 10.5755/j01.ppaa.20.1.28310

Podreka, Jasna, Gaber, Milica Antić, Pihler Ciglič, Barbara, & Kenda, Jana (2021). *Strokovne smernice: Preprečevanje spolnega in drugega nadlegovanja, trpinčenja in nasilja* [Professional guidelines: Prevention of sexual and other types of harassment, bullying and violence]. Ljubljana: Filozofska fakulteta, Univerza v Ljubljani.

Prpić, Katarina (1982). Žena u znanosti [A woman in science], *Žena*, 40(4), 53–67.

Prpić, Katarina (1989). *Marginalne grupe u znanosti* [Marginal groups in science]. Zagreb: Radna zajednica republičke konferencije Saveza socijalističke omladine Hrvatske; Institut za društvena istraživanja Sveučilišta u Zagrebu.

ReRIS11-20. *Resolucija o raziskovalni in inovacijski strategiji Slovenije 2011–2020* [Resolution on research and development strategy of Slovenia 2011–2020]. National Assembly [Državni zbor], Republic of Slovenia. Retrieved from http:// pisrs.si/Pis.web/pregledPredpisa?id=RESO68 [20.10.2021].

Rules on Academic Research Titles [*Pravilnik o raziskovalnih nazivih*], *Uradni list Republike Slovenije*, no. 126/08, 41/09, 55/11, 80/12, 4/13 – amended, 5/17, 31/17 & 7/19. Retrieved from http://pisrs.si/Pis.web/pregledPredpisa?id=PRAV9169 [20.10.2021].

Rules on the Procedures of (Co)financing, Evaluation and Monitoring of Research Activities [*Pravilnik o postopkih (so)financiranja in ocenjevanja ter spremljanju izvajanja raziskovalne dejavnosti*], *Uradni list Republike Slovenije*, no. 52/16, 79/17, 65/19, 78/20 & 145/20. Retrieved from http://www.pisrs.si/Pis.web/pregledPredpisa? id=PRAV12770 [20.10.2021].

Škerl Kramberger, Uroš (2017). Znanstveni šefi za mentorje predlagajo sami sebe, šefinj pa je tako in tako premalo [Science lords propose themselves as mentors, while not enough female counterparts]. *Dnevnik*, 26 June 2017. Retrieved from https://www.dnevnik.si/1042776274/slovenija/znanstveni-sefi-za-mentorje-predlagajo-sami-sebe-sefinj-pa-je-tako-in-tako-premalo [20.10.2021].

Slovenian ERA Roadmap. 2016–2020. *Slovenska strategija krepitve Evropskega raziskovalnega prostora 2016–2020* [Slovenian strategy for strengthening

European Research Area 2016–2020]. Ljubljana: Ministsrtvo za izobraževanje znanost in šport.

SPIRIT Slovenia. (2020). Slovenia. Green. Creative. Smart. (video). Retrieved from https://www.youtube.com/watch?v=CJIqTBV3iMs [20.10.2021].

Tomšič, Vida (1980) [1971]. *Ženska v razvoju socialistične samoupravne Jugoslavije* [A woman in development of socialist self-managing Yugoslavia]. Ljubljana: Delavska enotnost; Naša žena.

Ule, Mirjana, Šribar, Renata, & Umek Venturini, Andreja (Eds.). (2013). *Gendering science: Slovenian surveys and studies in the EU paradigms.* Wien: Echoraum.

Ule, Mirjana (2012). Spolne razlike v delovnih in kariernih pogojih znanstvenega dela v Sloveniji [Gender differences in research working environment and career progress in Slovenia]. *Teorija in praksa*, 49(4–5), 626–44. Retrieved from https://repozitorij.uni-lj.si/IzpisGradiva.php?id=85307&lang=slv [20.10.2021].

Ule, Mirjana (2013). Prikrita diskriminacija žensk v znanosti [Covert discrimination of women]. *Teorija in praksa*, 50(3–4), 469–81. Retrieved from http://www.dlib.si/details/URN:NBN:SI:DOC-9CU63AVL [20.10.2021].

Umek Venturini, Andreja. (2015). Women and science policy in the EU and Slovenia. In Mirjana Ule, Renata Šribar, & Andreja Umek Venturini (Eds.), *Gendering science: Slovenian surveys and studies in the EU paradigms* (pp. 155–60). Wien: Echoraum.

7 Disciplinary Communities of Practice for a greater gender equality in physics & life sciences

Sonja Reiland, Rachel Palmén and Lisa Kamlade

Introduction

The ACT project[1] has supported eight very different Communities of Practice (CoPs) to foster gender equality in research and innovation (R&I) and the majority of these CoPs have been either regionally based or thematically organised. Two of the CoPs however, were disciplinary based: the Life Sciences CoP and the GENERA CoP, which focused on physics. Whilst the aim of all CoPs was to some degree to work together to promote institutional change to further gender equality in R&I, the different focal points of each CoP have provided a rich source of experience about what works well and what does not work well, in collaborative, inter-organisational attempts to foster gender equality in R&I organisations. Whilst, it was impossible to directly compare the experiences of the regionally and thematically based CoPs with those that have taken a more disciplinary approach – this chapter aims to document the experiences of these two CoPs particularly looking at how they have been able to advance gender equality in their members' institutions – whilst reflecting on the advantages and disadvantages of taking a disciplinary approach to CoPs for institutional change.

In the following section, we will briefly describe the two CoPs that this chapter will discuss, regarding the member institutions, the shared vision and the basic organisational framework.

The GENERA CoP "Gender Equality in physics and beyond" originated from the EU-funded GENERA project (2015–2018) and its' vision is to support, coordinate and improve gender equality policies in physics research organisations in Europe and world-wide. A growing number of institutions joined forces to collaborate on institutional change. Currently, 40 Research Performing Organisations (RPOs), Research Funding Organisations (RFOs), and Higher Education (HE) member institutions are working together on the sex- and gender dimension in physics, career development for early career researcher, data collection, sustainability and outreach activities. Online meetings for the GENERA CoP happen on a monthly basis. Twice a year, the GENERA CoP meets face-to-face to set and monitor the

DOI: 10.4324/9781003225546-7

yearly defined agenda. Additional meetings take place within the Working Groups (WGs) dedicated to the action points for each year.

The main objective of the ACT – LifeSciCoP is to find practical solutions to change institutional culture towards gender equality. The Life Science CoP – builds on the work carried out in the European Union funded LIBRA project (2015–2019). The members of the CoP identified various topics they would be interested in working on, reflecting the whole "ecosystem" of gender-based discriminations. Nevertheless, the group agreed to tackle what they identified as the main bottleneck, i.e., systemic and personal gender biases, which are also reflected in the evaluation processes of researchers. The 15 partners of the LifeSciCoP are European research centres and university departments with a focus on life sciences. The professional roles of the individual members are very diverse, they occupy strategic positions such as head of human resources and director of operations as well as more implementation-based roles, like equality officers and principal investigators. In practical terms, the CoP members agree on specific actions and coordinate the work in individual WGs. The concrete topics the CoP is currently working on are diversifying institutional change agents (such as the gender equality committees), increasing institutional commitment, and providing guidance on institutional policy implementation and follow-up.

This chapter briefly identifies the relevant literature for considering a disciplinary based CoP for institutional change towards a greater GE in R&I and HE to frame the experiences and main findings of our two disciplinary based CoPs. It then discusses the methodological approach followed by the main findings which are structured by the following three concepts, domain, community, and practice. We then present some concluding reflections.

Literature review: Disciplinary-based CoP for gender equality in R&I

Discipline matters for gender equality in higher education, R&I in Europe and beyond. For example, the latest edition of She Figures 2021 highlights how the proportion of women among doctoral graduates varies according to fields of education. Women are over-represented in education (67%) but severely under-represented in the field of information and communication technologies (22%) and the fields of engineering and manufacturing and construction (29%). Career progression may also differ according to discipline, on average in the EU-27, in 2018, women represented 48% of doctoral graduates, which decreased the higher up the academic ladder – so to 47% of Grade C, 40% of Grade B, and 26% of Grade A. This gap however was wider in STEM – whilst women made up only 37.9% of doctoral graduates – less than 20% of Grade A academic positions were held by women (European Commission, 2021, p. 180). It has also been highlighted that available data across the broad STEM field could camouflage disciplinary specific causes of gender imbalances in career progression. There are many different facets to the relationship between career progression and disciplinary specific facilitators or obstacles for career progression. For example, the life sciences has been highlighted as one area where despite the fact that "women make up the majority of graduates up to doctoral level, they are less successful than men in obtaining research grants," especially in European Research Council (ERC) starting grants with 4.5% lower success rate (ERC, 2018, p. 57) whilst their numbers decrease the higher up the academic ladder (European Commission, 2021, p. 115). Interestingly, in Physical Science and Engineering, the report states that women have 0.9 higher success rate but make up only about 25% of applicants.

The women to men ratio of authorship and the percentage of scientific publications that integrate the sex and gender dimension also varies according to discipline. Within the pool of authors actively publishing, the number of men authors exceeded the number of women authors at all levels between 2015–2019 at both the European and country levels (European Commission, 2021, p. 8). She figures (2021) highlights how when the data is disaggregated by R&D fields, "gender gaps in active authorship are particularly prominent in the fields of Natural Sciences and Engineering and Technology" (European Commission, 2021, p. 8). The fields of Medical Health Sciences and Agricultural and Veterinary Sciences boasted the highest ratio of women to men authors – larger than 1.0 for both early-stage and middle-stage authors at the European level (European Commission, 2021, pp. 218–219). The integration of the gender dimension was most likely to be found in publications in the field of Medical and Health Sciences whilst publications on Engineering and Technology were least likely, followed by Natural Sciences (European Commission, 2021, p. 262).

Research has also demonstrated how disciplinary differences are important factors that must be taken into consideration for the successful implementation of gender equality interventions in R&I (Caprile et al., 2011). The European Commission has funded a raft of institutional change projects whereby consortias of between approximately 6 and 12 institutions/organisations from all over Europe and beyond come together to design, implement, monitor, and evaluate gender equality plans in R&I. Some of these projects are geographically-based TARGET, in the Mediterranean Basin, some of these are disciplinary based GENERA with a focus on physics, LIBRA with a focus on life sciences, or Equalist with a focus on ICT, or a combination of both, for example, Baltic Gender based in the Baltic Sea Region with a focus on Marine Science and Technology. This chapter builds on this work. Two of the CoPs supported by the ACT project were set up with a disciplinary focus: LifeSciCoP and the GENERA CoP. These were CoPs established to give some sustainability to the gender equality work already carried out through the LIBRA and GENERA Horizon 2020 projects. While the GENERA CoP included the majority of the project consortia members, LIBRA's project coordinator mainly brought together those institutions that were engaged with the project's dissemination activities (such as hands-on GEP workshops) and those recruited through the ACT coordinated stakeholder mapping for the LifeSciCoP. This chapter therefore aims to provide some key insights into the functioning of these two CoPs, regarding their domains, their communities, and finally their practice.

Examining CoPs as a vehicle to promote gender equality in higher education and research from a disciplinary approach is interesting for a variety of different reasons: CoPs tend to arise (emerge or are cultivated) in settings where knowledge is conceived as developed by practice – this is congruent with our aim of developing a CoP of gender equality practitioners in HE and R&I but may be seen as "contradictory" in a setting where knowledge production is "formalised" i.e., qualified and quantified through the production of "scientific" outputs, namely publications. What is considered knowledge and who produces knowledge are interesting questions in the context of knowledge producing institutions that aim to further gender equality. Researchers (formal knowledge producers), often both natural scientists and social scientists, come together with practitioners (gender equality practitioners) to design, implement, and evaluate gender equality measures aiming for institutional change.

Whilst CoPs have been studied in higher education (Hezemans & Ritzen, 2005; Jakovljevic, 2013), few studies have looked at the role of CoPs in advancing gender equality in research institutions (see Barnard et al., 2016). CoPs have been predominantly conceptualised as a vehicle for change within a research institution/organisation and even fewer studies have looked at how inter-organisational CoPs can advance gender equality within research organisations (Barnard et al., 2016; Thomson et al., 2021). The ACT project supported eight different CoPs, some thematically based, some regionally based, and two CoPs which were disciplinary based which is the focus of this chapter. So what issues are specific to disciplinary based CoPs promoting gender equality and gender mainstreaming in HE and R&I institutions?

CoPs work at the level of practice. So, those CoPs that function outside the realm of higher education – predominantly in the private sector see CoP members developing their "craft" through an apprentice form of learning – one could argue that whilst it is not "disciplinary" specific (in the academic field of science term) – it is practice specific – so for example early work on CoPs included ethnography of xerox workers or car manufacturer workers (McDonald & Cater-Steel, 2017, p. vi). These were CoPs that were established to improve the practice of workers through peer-to-peer learning. In the case of promoting gender equality in RPOs – this focus on practice highlights the relevance of gender equality practitioners. Interestingly, in the field of higher education, CoPs have been established within institutions to innovate in pedagogy (Maher, 2019). "Disciplinary practices" have, however, been highlighted as spreading beyond the realm of the specific institution fostering "disciplinary" collaboration (McDonald and Cater-Steel, 2017, p. vii) beyond institutional boundaries. However, the majority of CoP literature in higher education speaks of CoPs enabling an "interdisciplinary" approach – across boundaries of discipline (Fraser et al., 2017.; Kensington-Miller, 2017).

Morton, (2012) looks at a CoP in higher education – and highlights those disciplinary challenges associated with architecture. Lave & Wenger, (1991) saw that the success or failure in learning were characterised by mutual engagement, and therefore of great importance to them were those conditions that facilitated joint participation. Morton (2012) highlights how a "shared language" is a key resource for joint participation – which could be argued is aided by taking a disciplinary focus to inter-organisational CoPs. This may be particularly true for disciplines in the natural sciences like physics or life sciences where a specific vocabulary has been developed – so the sharing of this vocabulary can be seen to ease communication and understanding. Academic career development has also proven to be linked to broader disciplinary fields (e.g., STEM) – this is particular relevant for gender equality in academia and R&I (EC, 2019). Disciplinary CoPs for gender equality can work on tailored measures to ensure that disciplinary specific gender biases can be tackled whilst fostering professional development measures. Blanton & Stylianou in their 2009 paper "chart some directions for professional development that purposely use the content of a discipline to leverage issues of practice, recognising that the faculty will not be able to meet the goals of reform without the support to help them deal with the challenges presented…within their discipline in new and unfamiliar ways." (Blanton & Stylianou, 2009, p. 80). Hanrahan et al (2001) develops this line of thinking and states that "professional development that is discipline-specific and located in a community-of-practice is more likely to be relevant and productive than a centralized, decontextualized approach." Trends in the research systems are also often triggered by individual disciplines e.g., early adopters of pre-print publications are Physics and Economics, followed by Mathematics, and more recently by Computer Science and Biology (Morton, 2012). In terms of knowledge production, integrating the gender dimension into research content is structured, organised, and presented according to disciplines (European Commission, 2020).

Methodology

After identifying a gap in the literature looking at disciplinary CoPs for gender equality and institutional change, we developed the following research questions:

- What are the advantages and disadvantages of a disciplinary CoP approach?
- How should advantages be maximised and how can we overcome disadvantages?
- What are the similarities and differences between the two disciplinary CoPs?

- To what extent can the successes of the COP work be attributed to past initiatives?
- What enables joint learning?
- Does disciplinary homogeneity enable a more effective Community of Practice?
- How has collaboration in the CoP been aided by taking a disciplinary focus?
- How has collaboration in the CoP been hampered by taking a disciplinary focus?
- What are the main learnings and recommendations that we can take from physics and life science CoPs?

This chapter is based on a range of methodological approaches including a brief literature review, participant observation, as two of the authors facilitated the Life Science and GENERA CoPs and 10 semi-structured interviews with CoP members. The interviews were conducted by the CoP facilitators (n = 2) whereas the LifeSciCoP facilitator conducted the interviews with GENERA CoP members (n = 5) and the GENERA CoP facilitator conducted the interviews with the LifeSciCoP members (n = 5). The GENERA CoP facilitator has a background in business administration and business psychology and works as a project manager at a physics institute in Germany. The LifeSciCoP facilitator holds a doctoral degree in biochemistry and works in Spain as a senior scientific project manager.

The interviews took place in May 2021 and were conducted via video call (zoom). Each interview was approximately 30–45 minutes long. The CoP members were selected to create a diverse picture in terms of their gender, country, etc. However, of the interviewees chosen, only 30% were men since the majority of CoP members are women. The interview questions targeted the personal background of the CoP members and their perspectives on the benefits of the disciplinary focus of the CoP. The above research questions were operationalised into the following questions in the semi-structured interview guide:

1 Please explain a little a bit about your job role.
2 Please explain how you came to be involved in the GENERA/Life Sciences CoP.
3 How long have you been involved in this collaboration?
4 Can you explain a little bit about how the CoP works?
5 What benefits have you gained from your participation in the CoP? (Specific advantage of CoP approach?)

   a To what extent do you think that the disciplinary focus has been an important element in the functioning of the CoP and its relevance for you?
   b What other factors have facilitated the functioning of the CoP?

6 How has the CoP helped in your day-to-day work?
7 How has the CoP helped you to promote gender equality in your institution?
8 From your point of view, how could the CoP be improved?

Interviews were recorded and transcribed. Max QDA was then used to code the interview data. All authors independently coded some of the material in order to agree on the basic codes.

Main findings

The collected qualitative data were clustered according to the three dimensions of a CoP: domain, community, and practice, and we chart the benefits and challenges of choosing a disciplinary focus for a CoP throughout these three dimensions.

Domain

The domains of the LifeSciCoP and GENERA CoP – place different emphases on the three-gender equality and mainstreaming ERA objectives (representation of women in research careers, gender balance in decision-making and integrating the gender dimension into research content). The fact that the members of the two CoPs belong to a specific research discipline can be seen to influence the domain two-fold. First, the specific objectives of the CoPs can be discipline specific (e.g., attracting more women into physics or decision-making positions in the Life Sciences), and second, the CoPs' output and achievements are discipline specific and have the potential to have a greater impact not only on the stakeholders of the same discipline but also on the knowledge produced itself (e.g., guidelines for including gender dimension in research). Whilst both the Life Science and GENERA CoPs – placed most emphasis on creating institutional measures to deal with the first two objectives – GENERA did look at integrating the gender dimension into the research content of physics.

Diverging viewpoints existed as to the extent to which gender equality challenges pertained specifically to the discipline or could be linked to STEM disciplines more generally. One CoP member states, "we do not discuss science actually. We discuss the policies of the institutes and my institute is mainly an institute of life sciences but the problems are the same," whilst another LifeSciCoP member states:

> I think it is useful because the gender balance situation is very difficult.... Different in different sciences and for example, in bio-science there are a lot of women and at the lower levels it is completely women heavy. In our institute for example. 70% of employees are women... But that is very specific to Life Science and that would be very different if we were mixed with chemists, physicists where the situation is quite different.
>
> *(LifeSciCoP Member 2)*

This feeling was echoed by a GENERA CoP member as part of the reason to join the GENERA CoP in the first place, "it just sort of fascinated me... Physics which I think will be one of the toughest nuts to crack."

(GENERA CoP Member 3)

It is claimed that the life sciences is a particularly competitive research discipline, "hyped as the leading sciences of the 21st century, the life sciences have been very successful in attracting public money in recent decades. However, the almost explosive growth of third party-funded pre- and post-doctoral temporary job opportunities has not been matched by the number of new faculty positions (Stephan 2012)." (Fochler, 2016). Increased competition has been bolstered by the evaluation system of researchers mainly using quantitative bibliometric indicators. The research culture then adapts to this evaluation system, which encourages competitive behaviour, undermines open collaborations, and penalises researchers who commit to community services (such as institutional committee work, caring for, and teaching of students, etc.). The LifeSciCoP members shared the interest in changing this culture, which is reflected in one of its main objectives in developing guidelines for considering gender aspects in evaluating research performance. The presumed "meritocratic" approach in life science as well as in other disciplines has strengthened inequalities and disadvantages, especially for women who usually take over caring responsibilities in both professional and private life. An overall change in how research is evaluated including the consideration of gender aspects would definitely benefit from being piloted in an individual discipline, supported by a CoP.

In the GENERA CoP, integrating the gender dimension into physics was seen as particularly interesting:

I think we are really focusing, which I find very interesting ... in asking the hard questions in Physics ... we realised that we need to think a little bit wider because our borders between different subjects are artificial. so we usually say the maths intensive fields...and we talk about exploring the maths dimensions there. So it's like when we look at the gender innovation project for example, and then we realised that there is a ... a wealth of information, and very interesting when it comes to any topic that has a natural sex agenda dimension but very little on when it does not...and therefore that is what we are focusing on.

(GENERA CoP Member 1).

In the wide areas of physics and other math-intensive research, there is currently no accepted idea of how a diversity and gender perspective can be utilised. Resistance to incorporating a gendered perspective is often formulated in a form of the Haraway "God trick" argument (Haraway, 1988). In physics, this refers to the "lack of sex/or gender" in what is observed, i.e., planets and particles do not have sex or gender, or mathematics is only what is calculated: numbers, figures and formulas again do not have sex or gender (Genera CoP). The GENERA CoP decided to organise a conference to discuss a starting point in tackling this major challenge and to help define a convincing approach to show how a diversity and gender perspective must be present in these fields of science through teaching, defining research topics, performing research and its applications.

Community

The disciplinary domain is closely linked with the community dimension of the CoPs. In our work, this was highlighted by three different phenomena. Firstly, how the disciplinary nature of collaboration, built on existing projects and networks – defined the community. Secondly, how this community – crucially provided a safe space for those working on gender equality. Thirdly, the need to include men into gender equality work – this was seen to be particularly important in those disciplines where men are severely over-represented, like physics.

Due to disciplinary context members knew each other before they joined the CoP, either on a personal level, or on an institutional level, by meeting at conferences, or working together on collaborative projects and initiatives. In GENERA and in LifeSciCoP, former members of the H2020 projects GENERA and LIBRA continued their collaboration in the newly created CoPs. This provided the opportunity to sustain the collaboration beyond project partners. Additional institutions who followed the projects before in different ways (e.g., advisory board members, participants of workshops, etc.) or those who were less connected and were looking to engage with European-wide gender equality initiatives (e.g., identified through ACT Community Mapping) also became members of the disciplinary CoPs.

Research is an international endeavour which depends on collaborations and knowledge exchange, whilst the degree of cross institution and cross-country collaborations varies by institutions and disciplines. A disciplinary CoP can build on existing scientific or professional networks and on personal relationships in the ecosystem to strengthen also gender equality efforts. Existing connections between institutions through disciplines, even between people who are not representatives in the CoP gives accountability and leverage to the CoP. Focusing on the disciplinary dimension of a European CoP is one way of creating a feeling of belonging and understanding despite the diversity of nationalities, and countries of residence from members.

The "safe haven" in the CoP is created by the shared knowledge and experiences in the discipline specific context. CoPs provide precisely the means for establishing collegial relations in a safe place that is free of hierarchical power and politics typically observed in schools and faculties. Support from GENERA CoP was highlighted as a main benefit of belonging to the CoP:

> if you work on this inside the physics department or even in the science faculty you become very vulnerable. We in the university we change leadership unfortunately every three or six years or something and yes the swap can be that...suddenly, [you].have no support or are very isolated so a lot of the discussion is also I think pure therapy... keeping each other up. I think its very important to pick people up and also listen to the stories and what is happening.
>
> *(GENRA CoP Member 2)*

This support function of the community – proved a powerful mechanism – to enable the gender equality work of often isolated (in their own institutions) gender equality practitioners or lone scholars pushing for gender equality. In our CoPs, peers are defined as professionals who understand existing hierarchies and cultures in the represented organisations, which is often specific to disciplines, and who have either an intrinsic motivation to work on gender equality, or who have the institutional mandate to do so. This environment also attracts scientists belonging to disciplinary fields to engage with the CoP, as they feel qualified – even if they are not experts, nor are particularly experienced in the gender equality domain.

GENERA CoP members also stressed the need to create in inclusive CoP by including men. One of the main challenges was identified as follows:

> Making sure to get men on board because it is the same with Athena Swan ... you cannot have women doing all the work ... it is just not fair ... if you are really serious you have got to make sure that you have men who are getting involved as well, and they are shouldering the work ... I think it shows that it is being taken seriously at the discipline level and it sends a signal.
>
> *(GENERA CoP Member 3)*

Practice

Analysing the disciplinary aspects in the practice dimension of GENERA and LifeSciCoP, interviewees repeatedly highlighted the importance of the European dimension in terms of sharing good institutional practices. While regional or national focus can help to enhance the conditions for improving gender equality through lobbying on a policy and political level, the European dimension provides a platform to learn from different national contexts and to benefit from partners' experiences in countries with supportive legislative framework conditions. Sharing good practices is one effective way that inter-organisational CoPs can effect change at the institutional level. Since gender inequalities are often discussed on a disciplinary level – the disciplinary CoPs can provide the umbrella and practical tools for collecting benchmarking data with the goal to inform institutional strategies, priority developments, and relevant policy makers. One of the working groups within GENERA CoP focuses on developing teaching and training materials based on good practices from the participating institutions. The aim is to design and implement a workshop on career development issues for early career researchers bringing in a gender and diversity perspective. On another level, GENERA helps other institutions to create their own gender equality plan (GEP) by sharing its best practices and knowledge from members. One of the tools used is the toolbox which was already developed during the GENERA project and is now assisting organisations that are in the process of the implementation of GEPs in tailoring their GEPs and gender equality measures to their needs. For this purpose, the toolbox offers a range of measures that can serve as models for other organisations.

The LifeSciCoP has the objective of exchanging good practices about the evaluation of researchers. Changing practices need to be done on a community level, rather than at the institutional level, as researchers tend to be very mobile during their careers, changing institutions. In terms of scaling up good practices for institutional change – often gender equality measures are started at the institutional level and individual departments/schools adopt measures or modify according to their specific context. Nevertheless, in some cases, it can be the reverse (expansion of the disciplinary dimension to whole institution) or in parallel (parallel institutional and unit/disciplinary level).

A CoP at the European level can be seen as an "insurance" measure to not let gender equality depend on the good will of national government whilst providing a supportive "infrastructure" to develop its' priority at the institutional level, whilst sharing institutional good practices. CoPs brought together institutions with a long trajectory of gender equality work with those that were newer to this field – this was seen to provide a really useful function that ensured efforts were streamlined.

> I think in our CoP what we feel most...particularly the groups where some were not very advanced at all with equality issues- not having a [GEP], not having things established is that it has brought tremendous strength and brotherly and sisterly spirit in. knowing that other people are in the same boat has been...everybody says it is really important and learning from other people. It's ... it's a huge source of strength and people are realising that they do not have to reinvent the wheel. They can take best practices from other people, they can reuse activities, reuse .... And that is really important because ... everybody likes to invent the wheel and they do invent the wheel and a lot of time and effort is wasted.
>
> *(LifeSciCoP Member 4).*

In the LifeSciCoP, a very clear example was given about how institutional measures can be shared between institutions operating in very different contexts, and how those with a more developed legislative and policy framework can positively impact on institutions operating in contexts with less developed policy and legislative frameworks. As a CoP Member explains:

> my colleague ... was asking about prevention of harassment protocols because in X she said they do not have much documentation in their institution which is quite large. They have a person responsible for this, but they do not believe that person has had any training ... did we have any guidelines? Well we have quite strict guidelines ... X law ... law in X tends to be very prescriptive. So they have exactly what you want to have which makes it quite easy to prepare because you are told what you have to have. So we have a document that is usable and we are still preparing it but that is something I can share with her and in the absence of anything else she can adapt it quite directly I think ... In the CoP what we are doing now is collecting documentation on policy guidelines, things that we can use, that people can pass around.
>
> *(LifeSciCoP Member 2).*

Conclusions

We have documented the extent to which organising a disciplinary CoP for advancing gender equality in R&I has been a useful strategy through producing knowledge about institutional change and integrating the gender dimension in research content (domain), through defining the CoP members – including stressing the importance of involving men and the support mechanisms of this approach (community) and through what they do – particularly sharing good practices (practice).

Regarding the domain there was no consensus as to the extent to which taking a disciplinary focus was beneficial. Even if life sciences is the CoPs domain – the discipline was not very present in the discussion. It was rather the assumed context that the members agree on. In the GENERA CoP however, a disciplinary approach favoured integrating the gender dimension into research content, i.e., integrating the gender dimension into physics was seen as a potential area for real discussion, interest and growth – as very little work had been done in this field to date. GENERA CoP is currently organising one of the first spaces to discuss integrating the gender dimension into research content and have begun to lay the foundations for this interesting contribution to knowledge.

Regarding the community – it has been highlighted how disciplinary CoPs can be valuable to catalyse innovation and progress within disciplines across geographic and cultural boundaries (MacGillivray, 2017, p. 42). Our experiences and research carried out in the framework of the ACT project highlights how disciplinary CoPs have been able to catalyse innovation and progress for gender equality in R&I at the institutional level – across geographical and cultural boundaries. The transnational European disciplinary community (physics or life sciences) and networks provided a good basis for collaboration on gender equality in R&I organisations. The community has provided a safe-space and support – infrastructure for gender equality practitioners to share experiences and offload. This has proved invaluable – and is perhaps one of the most important yet least tangible outcomes of the CoP approach. This however is not disciplinary specific. The call to include more men into shouldering the workload for gender equality – has to be welcomed and came from the disciplinary CoP where men are severely over-represented.

In terms of practice, interviewees appreciated learning from peers by talking and exchanging, which is comparable to the xerox workers or car manufacturer examples discussed in the literature review (McDonald & Cater-Steel, 2017, p. vi). The organisation of work in the CoPs – through working groups, and overall CoP meetings – as goal oriented with medium- and long-term objectives was seen as key. Sharing and exchanging good practices – was seen as the real motor driving forward gender equality in each member institution.

Specifically forming part of a European project was seen as giving legitimacy to the CoP, "membership of being part of a European project – that's very important because as you know scientists take these things into account. It is not just a vague voluntary thing – it is an organised … an organised project with tangible objectives so that is important." (GENERA CoP Member 5). The funding was also seen as key as it provided resources for a CoP facilitator that was seen to be crucial to the smooth functioning of the CoP – whilst resources were also seen to be key to institutional change.

Note

1. "Communities of PrACTice for Accelerating Gender Equality and Institutional Change in Research and Innovation across Europe" Horizon 2020 project, grant number 788204 is referred to throughout this book as "The ACT project". See also https://www.act-on-gender.eu.

References

Barnard, Sarah, Hassan, Tarek, Dainty, Andrew, Polo, Lucia, Arrizabalaga, Ezekiela. (2016). Using communities of practice to support the implementation of gender equality plans: Lessons from a cross-national action research project. Loughborough University. Conference contribution. https://hdl.handle.net/2134/23681

Blanton, Maria, & Stylianou, Despina. (2009). Interpreting a community of practice perspective in discipline-specific professional development in higher education. *Innovative Higher Education, 34*(2): 79–92.

Caprile, Maria, Addis, Elisabetta, Castaño, Cecilia Klinge, Ineke, Larios, Marina, Meulders, Danièle, Müller, Jörg, O'Dorchai, Síle, Palasik, Maria, Plasman, Robert, Roivas, Seppo, Sagebiel, Felizitas, Schiebinger, Londa, Vallès, Nuria and Vazquez-Cupeiro, Susanna. (2012). Meta-Analysis of Gender and Science Research, Publications Office of the European Union, Brussels.

European Commission, Directorate-General for Research and Innovation. (2019). She figures 2018, Publications Office of the European Union, Brussels. https://data.europa.eu/doi/10.2777/936

European Commission, Directorate-General for Research and Innovation. (2020). Gendered innovations 2: How inclusive analysis contributes to research and innovation: Policy review, Publications Office of the European Union, Brussels. https://data.europa.eu/doi/10.2777/53572

European Commission, Directorate-General for Research and Innovation. (2021). She figures 2021: gender in research and innovation: Statistics and indicators, Publications Office of the European Union, Brussels. https://data.europa.eu/doi/10.2777/06090

European Research Council (ERC) Annual Report on the ERC activities and achievements in 2017. Publications Office of the European Union, Luxembourg.

Fochler, M. (2016). Variants of epistemic capitalism: Knowledge production and the accumulation of worth in commercial biotechnology and the academic Life Sciences, *Science, Technology and Human Values, 41*(5): 922–948.

Fraser, Cath, Honeyfield, Judith & Breen, Fiona. (2017). 'From project to permanence: Growing inter-institutional collaborative teams into long-term, sustainable Communities of Practice'. In *Communities of Practice. Facilitating Social Learning in Higher Education*, edited by Jacquie McDonald and Aileen Cater-Steel. Springer.

Hanrahan, Mary, Ryan, Micheal, & Duncan, Margot. (2001). The professional engagement model of academic induction into online teaching. *International Journal for Academic Development, 6*(2): 130–149.

Haraway, Donna. (1988). Situated knowledges: The science question in feminism and the privilege of partial perspective. *Feminist Studies, 14*(3): 575–599.

Hezemans Marijke, Ritzen Magda. (2005). Communities of Practice in higher education. In: Tom, van Weert, & Arthur Tatnall. (eds), *Information and Communication Technologies and Real-Life Learning. IFIP – The International Federation for Information Processing, 182*. Springer. https://doi.org/10.1007/0-387-25997-X_5

Jacquie, McDonald and Aileen, Cater-Steel. (2017). Communities of Practice. *Facilitating Social Learning in Higher Education*. Springer. Accessed 11 October 2018.

Jakovljevic, Maria., Buckley, Shirley., & Bushney, Melanie. (2013). Forming Communities of Practice in Higher Education: A Theoretical Perspective. Active Citizenship by Knowledge Management & Innovation, 19–21 June, 2013, Zadar, Croatia.

Kensington-Miller, Barbara. (2017). Surviving the first year: New academics flourishing in a multidisciplinary community of practice with peer mentoring. *Professional Development in Education, 44*(5):1–12, https://doi.org/10.1080/19415257.2017.1387867

Lave, Jean, and Etienne Wenger. (1991). *Situated Learning: Legitimate Peripheral Participation*. Cambridge University Press.

MacGillivray, Alice. (2017). "Social learning in higher education: a clash of cultures?," in McDonald, J. and Cater-Steel, A. (eds), *Implementing Communities of Practice in Higher Education*, Springer, pp. 27–45.

Maher, Damian. (2019). "The use of course management systems in pre-service teacher education," in (Ed.) Jared Keengwe, (eds), *The Handbook of Research on Blended Learning Pedagogies and Professional Development in Higher Education*, IGI Global, pp. 196–213.

Morton, Jane. (2012). Communities of practice in higher education: A challenge from the discipline of architecture. *Linguistics and Education*, *23*(1), https://doi.org/10.1016/j.linged.2011.04.002

Stephan, Paula. (2012). *How Economics Shapes Science*. Harvard University Press.

Thomson, Aleksandra, Palmén, Rachel, Reidl, Sybille, Barnard, Sarah, Beranek, Sarah, Dainty, Andy., & Hassan, Tarrek M. (2021). Fostering collaborative approaches to gender equality interventions in higher education and research: The case of transnational and multi-institutional communities of practice. *Journal of Gender Studies*, *0*(0), 1–19. https://doi.org/10.1080/09589236.2021.1935804

8 A gender budgeting Community of Practice and targeted implementation projects

A potential to challenge gender biases in decision-making in research organisations?

Laufey Axelsdóttir, Finnborg Salome Steinþórsdóttir and Þorgerður J. Einarsdóttir

Introduction

Confronting gender bias in decision-making has been defined by the European Research Area as one of three priorities for the advancement of gender equality in higher education, research, and innovation. Actions to eliminate gender bias in decision making have generally focused on correcting the gender imbalance in leadership of research performing organisations (RPOs, hereafter 'research organisations'). However, that alone is not a sufficient condition to eliminate gender biases in decision making. It has been documented that decision making processes, through which organisations formulate budgets and allocation of resources, maintain and continue to produce gender inequalities in research organisations. This can be reflected, for instance, in government's teaching rates per student, and research performance indicators that are more favourable to the male dominated and 'masculine' STEM disciplines than other more gender balanced fields (Steinþórsdóttir et al., 2019). Gender budgeting is a strategy that can be used to tackle gender bias in decision-making as it acknowledges that financial management mechanisms are not gender neutral (Addabbo et al. 2020; O'Hagan, 2018; Sawer and Stewart, 2020).

Gender budgeting is a relatively new strategy to advance gender equality and efficiency of policy making in the context of research organisations (Addabbo et al., 2020; Steinþórsdóttir et al., 2019). The authors of this chapter, who have the responsibility to support and mentor the Community of Practice (CoP) in their capacity as seed partners, have been researching and developing gender budgeting for research organisations as part of the GARCIA research project and the Gendersense network. In the GARCIA research project, funded by

DOI: 10.4324/9781003225546-8

the European Union Seventh Framework Programme in the years 2014–2017, the researchers were responsible for developing gender budgeting in research organisations (GARCIA, n.d.). The Gendersense network was established as a continuation of the GARCIA project and involves research organisations in Europe, North- and South America, the Icelandic Governmental Offices, and the City of Reykjavík. The objective of Gendersense is to research and enhance knowledge on inequality regimes in education (primary, secondary, and tertiary education). It serves the implementation of gender budgeting, as well as developing an inclusive and intersectional approach to the strategy, that we refer to as gender+ budgeting. ACT's first Synergy conference (ACT, 2019) provided an opportunity to explore interest on configuring a CoP on gender budgeting in research organisations and brainstorm on how to cultivate such a community open to members from all geographical regions. Knowledge or experience on gender budgeting was not a precondition for participating in the CoP, only a commitment to engage in and participate in developing gender budgeting to tackle gender biased outcomes of decision-making in research organisations.

The ACT GenBUDGET CoP was formed by 16 representatives from 10 research organisations in 2019. The objective is to develop shared knowledge, resources, and practices on how to engage effectively in the design, implementation, monitoring and evaluation of gender budgeting for institutional change in the participating organisations. To emphasise the aim of advancing gender equality within the research organisations, the seed partners encouraged the CoP members to take on, 'targeted implementation projects' (TIPs), at their own discretion, based on local conditions and circumstances at each member's institution. The experience from the GARCIA project indicates that a lack of transparency and gender-disaggregated data is a common problem in European universities. Hence, it was deemed effective to start with manageable projects in which data and statistics are available, and preferably, relating to some complementary projects (Steinþórsdóttir, Hejstra, Einarsdóttir, and Pétursdóttir, 2016). In their TIPs, the CoP members assess the gender impact of one or more financial managerial mechanisms they have chosen within their research organisation. Drawing on their findings the CoP members formulate measures that have the potential to advance gender equality, if implemented by the governing authorities in their organisations. The CoP is a venue for sharing the process and outcomes of the TIPs where the CoP members receive support and advice from the seed partners and the whole GenBUDGET community. Hence, through the TIPs the CoP members obtain in-depth knowledge on the 'inequality regimes' (Acker, 2006) in research organisations and expertise on how to integrate a gender dimension into the decision-making processes and ensure more gender-equal outcomes.

In this chapter, we use a case study approach to reflect on the opportunities and obstacles of the GenBUDGET CoP in developing and implementing gender budgeting to challenge gender biases in decision-making of research organisations. To provide insights into GenBUDGET and its progress, we build on Wenger, McDermott and Snyder (2002) theoretical writings on CoPs. Our analysis draws on the experiences and knowledge gained through the CoP's meetings, workshops, webinars, and blogs. We explore the potential to harness inter-institutional cooperation in adopting gender budgeting in an international CoP, when CoP members' knowledge about the strategy is at different place. Furthermore, we examine whether and how the CoP approach, by the means of TIPs, has enhanced knowledge on inequality regimes and supported the adoption of gender budgeting in their organisations.

CoP as a social learning system

The concept of CoP was developed by Lave and Wenger (1991). According to them, learning is situated in the trajectories of participations in the social world (Lave and Wenger, 1991), through interplay of social competence and personal experience. By participating in complex 'social learning systems', knowledge is created in a cultural and historical context (Wenger, 2000). A CoP can be observed as a 'social learning system' (Wenger, 2000), as it focuses on a domain of knowledge and developing shared practice (Wenger et al., 2002). CoPs can take different shapes and vary along several dimensions. While some are small and tight-knit, others are large and more loosely connected. However, they all share certain characteristics (Wenger, 2001). A CoP has been defined as: 'group of people who share a concern, a set of problems or a passion about a topic, and who deepen their knowledge and expertise in this area by interacting on an on-going basis' (Wenger et al., 2002, p. 4). To create knowledge, people need opportunities to engage with others in similar situations where they can join shared actions, thinking, and conversations. Therefore, the group meets as they find value in their interactions by sharing information, insight, and advice. Also, by helping each other solve problems, discuss their situation, aspiration, and needs. The interactions are important as they allow the community to develop a unique perspective on their topic and common knowledge, practice, and approaches (Wenger et al., 2002).

Wenger et al. (2002) developed a three-dimension model which is pertinent to the GenBUDGET CoP (Palmén et al., 2019). First dimension is shared interest in a domain of knowledge, it consists of key problems commonly experienced by the members, which create a common ground and a sense of common identity. The second element is community, expressing the need to engage in joint activities and discussions, help each other and share information to pursue their interest. To build a CoP, the members must interact on a regular basis on issues important to the domain. The third element is practice, wherein the community develops, shares, and maintains special knowledge. Moreover, there are also three main levels of community participation. First is a core group of people who actively participate in the CoP and are the heart of the community. Then there is the active group who attends meetings regularly and participates occasionally in the community's forum, although, not as intense as the core group. Finally, is the peripheral group that rarely participates, but watch the interaction of the core and active members (Wenger et al., 2002).

Gender budgeting

Gender budgeting is a structural transformation strategy that is used to advance gender equality. Gender budgeting efforts have been initiated in more than 80 countries around the world with different forms and approaches, as well in different settings (Budlender and Hewitt, 2002; Stotsky, 2016). The strategy is most commonly defined as an integration of the gender perspective in all aspects and all stages of the budgetary process to promote gender equality (Council of Europe, 2010). However, for gender budgeting to be more than a bureaucratic exercise, it requires a feminist approach. Drawing on O'Hagan (2018), we approach gender budgeting as a strategy that aims to challenge the supposed gender-neutral policies, programmes, and resources that re/produce gender biases and inequalities and change organisational processes to ensure that they promote gender equality. This approach directs the attention from the individuals to the structural dimensions that produce and maintain gender+ inequalities. Acker (2006) defines these practices and processes as 'inequality regimes'. Inequality regimes are in all work organisations but are various and changeable and connected to wider societal inequalities. The inequality regimes can differ in the degree of visibility (the awareness of inequalities) and legitimacy (the justification of the inequalities). High visibility and low legitimacy of inequality increases the potential to transform inequality regimes.

In the context of research organisations, gender budgeting is a relatively new strategy to facilitate gender equality, but research is scarce on how financial management mechanisms maintain and even produce inequality regimes (Acker, 2006). In our previous work on gender budgeting in research organisations (Steinþórsdóttir et al., 2016; 2017; 2018; 2019) we have utilised gender budgeting as a lens to evaluate financial and managerial processes and procedures. We have found that new managerial methods and instruments, such as performance indicators and incentive systems, create gendered structural hindrances that generate and foster gender inequality. With our work, we aim to surmount these institutional mechanisms to promote more gender equal outcomes. Other initiatives have focused on how finances can be used more broadly to encourage gender equality work in general within the academic system (Addabbo, Rodríguez-Modroño and Gálvez-Muños, 2015; Erbe, 2011, 2015; Rothe et al., 2008). Addabbo et al. (2020) mapping of gender budgeting methodologies in European research organisations found that the most common methodology used is gender mainstreaming of the public finance management processes, that is, integrating gender to the planning, implementation, auditing, and evaluation of the budget. This is followed by the account-based approach, that is, assessing the consistency between gender objectives and budget-allocations. Other approaches include well-being gender budgeting, that is, evaluating the contribution of the research organisation to the construction of people's well-being, and performance-based budgeting, linking gender equality targets to the budgeting. Regardless of the approach and the methods applied, O'Hagan (2018, p. 37) argues that 'gender budgeting requires a feminist approach [...] [that] seeks to force policymakers to consider, reveal, and rethink inbuilt gender biases in the processes of analysis and decision-making, and processes of engagement and participation'.

Methods

We employ a case study approach to conduct a thorough, longitudinal analysis of the progress of the GenBUDGET CoP in developing and implementing gender budgeting to challenge gender biases in decision-making of research organisations. A case study entails a detailed analysis of a specific case, such as a community or an organisation, and frequently includes a longitudinal element, where the researcher is a participant of the community or the organisation involved (Bryman, 2012), as is the case with the authors of this chapter. Empirical data was collected from May 2019 to May 2021. First, through a survey with nine questions (eight closed and one open), that the GenBUDGET seed partners sent out to the CoP members at the beginning of the CoP's operation. The intention was to map the initial knowledge state of the participating organisations in gender budgeting, and to find out if the organisations had a gender equality plan and gender equality officer. Second, through a documentation of 14 monthly meetings, two online workshops, and three webinars, and examination of nine blog posts from the CoP members on the ACT GenBUDGET (n.d.) website (published in between September 29, 2020, and June 6, 2021). This allowed us to conduct detailed examination of GenBUDGET CoP members' experiences and activities, including their successes and challenges.

One of the monthly meetings was a face-to-face meeting in Hamburg, Germany, in January 2020. The first workshop consisted of three meetings in April and May 2020 (2020 Workshop), and the second workshop included one meeting in April 2021 (2021 Workshop). The webinars were held in December 2020, February 2021, and May 2021. To support productive conversation between the CoP members in the meetings, the co-creation toolkit provided by the ACT project was used. The toolkit's methods aim to help CoPs operate, develop, implement gender equality plans, gender equality measures and activities, and facilitate institutional change in relation to GE (Thomson and Rabsch, 2021). Furthermore, we prepared an agenda prior to the meetings and events, with questions for the CoP members to reflect upon. This included questions such as: *How has your TIP progressed? What is your situation regarding TIPs and COVID-19? What support do you think the CoP needs to enhance knowledge and develop shared Gender Budgeting practices in research organisations? What is your experience of using the ACT community support measures to achieve gender equality objectives? What are the main benefits of the support activities already received? Are there some resources/ supports missing in your opinion? If so, which ones? What do you foresee as the CoPs next steps? How are the gender budgeting TIPs useful to achieve gender equality objectives? Do you foresee any actions taken to facilitate equality? Are TIPs a useful approach, or do we need to rethink the approach?*

The statistical package for the social sciences (SPSS) computer software (version 26) allowed us to use descriptive statistical measures to analyse the survey data. The findings were used to shed light on the CoP members shared interest of gender budgeting, thus, the domain. To analyse the data collected in the meetings, workshops, and webinars, we carefully read through all the documents we collected. In order to describe the meaning of the data in a systematic way, we applied a qualitative content analysis (Schreier, 2013), in which we emphasised assessing the possibilities of GenBUDGET for creating knowledge about gender budgeting. We also used the method to analyse how the CoP approach and the collaboration has supported gender budgeting in the participating organisations. We are aware of the limitations of the qualitative data collection and analysis. There is a certain risk that people do not say everything they think, and some find it difficult to express themselves in such a group. However, we think important lessons can be drawn from a systematic account and portrayal of the experiences of the CoP.

Findings

To analyse the data, we use a three-dimension model from Wenger et al. (2002), adapted by the ACT project (Palmén et al., 2019). In the following, we explore the GenBUDGET development and address how these dimensions are reflected in the CoP.

Domain: A shared interest in gender budgeting to progress gender equality in RPOs

The main domain of the GenBUDGET CoP is tackling gender bias in decision making processes by the means of gender budgeting. The emphasis is on the gendering managerial mechanisms through which organisations formulate budgets and allocation of resources. Wenger et al. (2002) point out that a critical mass of people is required to sustain regular interactions. The existence of shared practice is what allows CoP members to share knowledge and offer multiple perspectives. When GenBUDGET was established in 2019, the CoP had 16 representatives from eight universities and two research institutions. In January 2021, five representatives from four universities joined the CoP. Thus, the CoP has 21 representatives from 12 universities and two research institutions distributed over 11 countries, i.e., the Nordic countries, Western and Southern Europe (see Table 8.1). The

Table 8.1 Information about the CoP members

University	Research institute	Country	Number of employees	Number of students
University of Iceland		Iceland	4465	13092
University of Southern Denmark (SDU)		Denmark	3171	22558
Örebro University		Sweden	1600	15000
	Western Norway Research institute	Norway	30	0
Vilnius University Kaunas Faculty		Lithuania	3,095	22,747
University of Birmingham		United Kingdom	7,000	30,000
Glasgow Caledonian University		Scotland	1,600	16,860
RCSI – Royal College of Surgeons in Ireland		Ireland	1,160	4,094
University Carlos III of Madrid (UC3M)		Spain	2,081	22,666
	Fondazione Giacomo Brodolini (FGB)	Italy	49	0
Ulster University		Northern-Ireland	1,665	24,530
Open University		United Kingdom	8,242	168,167
RWTH Aachen University		Germany	9,826	47,173
University of Modena & Reggio Emilia (UNIMORE)		Italy		21,147

number of employees in these universities range from 1,160 (Royal College of Surgeons in Ireland) to about 9,826 (RWTH Aachen University). The research organisations, on the other hand, have 30 and 49 employees. The number of students in the universities varies, with the highest number in the Open University (about 168,167) and the lowest in RSCI (4,094 students). Despite the differences between the organisations, they share a common interest which is a strong motivation for joining a CoP and build shared practice. Wenger et al. (2002) point out that a CoP with less than 15 members can be considered as a very intimate community. With reference to that, our CoP is close to being that kind of community.

Membership in a CoP implies a commitment to the domain and a shared competence (Wenger-Trayner and Wenger-Trayner, 2015). All GenBUDGET members are gender experts and/or have worked in the field of gender equality through teaching and research, development projects, and administration. Thus, sharing an interest in gender theory and strategies to advance gender equality. Their experiences and knowledge of gender budgeting was documented in a survey during the setting-up of the CoP, revealing diversity of the organisations and dissimilar experience of the strategy. Three of the CoP members' organisations had participated in a research project on gender budgeting in research organisations, and three of them had either worked on gender budgeting in the context of a research organisation or at different governmental levels. Four of the participants had no previous experience of gender budgeting. Although, previous experience or knowledge on gender budgeting was not a precondition for participating in the CoP, it has been very valuable for the CoP. The CoP members have been able to share knowledge and give important insights into previous research projects on gender budgeting. Thus, the collaboration in GenBUDGET has been important for the CoP's development of shared knowledge. As Wenger et al. (2002) pointed out, in order to develop expertise, practitioners need opportunities to engage in a process of collective learning with others who face similar situations.

The survey findings also demonstrate that most of the participating organisations have a gender equality plan emphasising gender mainstreaming and employ gender equality officer/s. However, only one of the organisations has gender budgeting included in the gender equality plan, while two organisations are currently working on it. All the initial CoP members' organisations showed their dedication to advance gender equality and to work on gender budgeting by signing a Memorandum of Understanding (MoU) in the beginning of the ACT project.[1] To create more solidarity among the CoP members, support their work toward a common vision for community, and build trust and relationships, the seed partners drafted a 'Tailored Support Package', entailing the 'CoP's vision, mission and agenda. To ensure that GenBUDGET members had a say in this process, the support package was distributed to the CoP members to get their ideas and comments. Through this collective work, the support package was finalised and the understanding of the community's domain established. As Wenger et al. (2002) explained, it is the CoP members commitment to shared learning agenda that motivates them to contribute to the community.

Community: A growing international collaboration

The GenBUDGET CoP includes both practitioners and researchers, who communicate and meet regularly to promote gender equality in the decision-making area, in accordance with the CoP approach agreed on in ACT (Palmén et al., 2019). In the beginning, the seed partner had meetings with each CoP member to apprehend their needs and aspirations for GenBUDGET. Since September 2019, the GenBUDGET CoP members have engaged in monthly meetings to discuss their activities, approaches to gender impact assessment and their projects. Thus, the community is a site where members interact, share experiences and learn from each other, share their concerns and seek guidance on how to tackle hindrances, indifferences, and other problems that might occur in the implementation process. The meetings have mostly been online, apart from one face-to-face meeting in Hamburg, Germany, in January 2020, which the ACT project made possible. As learning requires an atmosphere of openness, where it is safe to speak the truth and ask hard questions (Wenger et al, 2002), the face-to-face meeting was of utmost importance for GenBUDGET's growth. Following, the CoP members seemed to experience more capacity to contact and help each other when needed. Their relationships seemed to be more relaxed than before, indicating that the meeting was an important opportunity for the CoP members to build trust and relationships, share ideas with other members, ask questions, and listen to each other. As Wenger et al. (2002) explained, learning is a matter of belonging. CoP members' commitment is also reflected in their participation in discussions. Majority of the CoP members could be categorised as core members, who actively participate in the CoP (Wenger et al., 2002).

In addition to the face-to-face meetings, the online workshops and webinars have also been important in establishing and sustaining GenBUDGET's sense of community. The 2020 workshop created an opportunity for the CoP to have interactive discussion on GenBUDGET's next steps and the sustainability of the community. The workshop revealed the need to organise a new venue where members with similar projects meet and discuss, stressing how important it is to recognise the CoP members' dissimilar contexts and projects to ensure their sense of belonging in the community. As Wenger et al. (2002) point out, CoP members develop a unique individual identity in relation to the community, wherein their interactions are source of both commonality and diversity. Therefore, four working groups have been established: one on research funds, another group focusing on the COVID-19 and its influences on research institutions, as well as a working group on the gender pay gap, and finally a group on workload allocation.

The objective of the smaller groups is to support the CoP to move forward in a focused and purposeful way.

Another outcome of the 2020 workshop's discussion, was an open webinar organised in December 2020 to allow the CoP members to present their ongoing projects, share experience, and continue the consensus building of gender budgeting practices. The CoP members were pleased with the webinar, as they thought it was helpful to get insights into the progress of the TIPs. Following, two other webinars were held, in February 2021 and in May 2021. Not only do the webinars reflect the CoP members work, but they are also a fruitful collaboration between the members.

In order to enable the collaboration between the CoP members, the ACT project provided access to experts in gender budgeting in monthly meetings, access to resources, and the co-creation toolkit on the Knowledge Sharing Hub (ACT GenBUDGET, n.d.). The CoP members learned to appreciate the co-creation toolkit and its various methods to support productive conversation when they met face-to-face in Hamburg. Moreover, in the 2020 workshop, the CoP members discussed how the toolkit has been helpful in real-life and virtual meetings, and one mentioned recommending the co-creation toolkit with a colleague. The tools help the CoP members find things in common, both positive elements and hindrances, thus creating a learning process. The CoP members have underlined that the support provided within GenBUDGET is important, some kind of 'peer consultation' so they do not have to reinvent the wheel. Having an opportunity to learn from the discussions among the CoP members and share experiences is valuable. The members are partly encountering similar obstacles, such as accessing data and resources, and getting through to decision-makers. They experience GenBUDGET as a social support, wherein they are not alone in their struggle for gender equality and they find it important to stay in contact, and think of the CoP as a way to energise. The CoP members have discussed how the CoP has been useful to make associations between members, to share what is going on in other countries and what can be done to implement gender budgeting in research organisations. The CoP opens opportunities to build coalitions at the national level. Moreover, good experience of the community's interactions is reflected in the CoP members' collaboration outside the CoP's activities. Two of the CoP members, for instance, participated in an online workshop organised by one of the members.

The COVID-19 pandemic has influenced the activities and development of the community at many levels, the most apparent being restrictions from further face-to-face meetings and workshops. The CoP had planned a workshop in Iceland in April 2020, which was cancelled due to the pandemic. Therefore, three online meetings were organised in April and May 2020. In the 2020 workshop, the CoP members discussed how the pandemic had created entirely new situations with increased workloads, abrupt shift to online teaching which was new to many of them, and increased student support, which for some was urgent and an extensive task. This affected their research heavily and gave them less time than they wanted to participate in the CoP. Some of the CoP members shared with the group how the pandemic has influenced their personal life as they are experiencing more family responsibility than before the pandemic, especially childcare and home schooling. However, although participating in the CoP's work may have been difficult at times, the CoP members also found support in learning from each other and sharing their experiences at the meetings.

The GenBUDGET experiences of collaborating in the CoP during the COVID-19 pandemic were addressed at the 2021 workshop, where the CoP members identified both threats as well as opportunities related to COVID-19. They fear that the pandemic has resulted in less resources and budget cuts, especially affecting topics of low priority such as gender equality, as well as less availability of gender-research funding. However, by exposing and making worse the already existing barriers to women's career progression and research outputs, COVID-19 has created an opportunity to make them visible and force a change in practice. O'Hagan (2021) pointed out that in order to remedy the impacts of COVID-19, actions should designate specific funds to support academics, mainly women, who have lost time to COVID-19. Despite the barriers created by the pandemic, the CoP has managed to stay active. In fact, it has grown in size, as mentioned earlier. Although not being able to meet face-to-face, the CoP members have been dedicated to participate in the monthly meetings and planned events during difficult societal reality, reflecting the intimacy of the CoP members who experience the encouragement discussed by Wenger et al. (2002) to actively take part in the community.

Practice: Developing shared knowledge on implementation of gender budgeting

Wenger et al. (2002) state that successful practice coincides with community building, wherein the process must give the CoP members reputations as contributors to the community's practice. By sharing experiences, stories, tools, and ways of addressing problems in the meetings, workshops, and webinars the GenBUDGET members have actively contributed to the community's practice. The diverse TIPs established by the CoP members played a key role, including projects on the distribution of financial funds, gender pay gap, workload allocation schemes, internal research grants processes, the status of sessional teachers, and postgraduation status of PhD earners.

As knowledge levels on gender budgeting and gender inequality was at different places in the participating organisations to begin with, it was challenging for the members to start their TIPs. Wenger et al. (2002) state that in order to handle new situations and create new knowledge, it is important to provide resources to enable that process. Various resources have been provided to reach this goal. A framework provided by the seed partner has supported the CoP members to find and establish suitable TIPs for their organisation. The GARCIA gender budgeting toolkit was made available on the ACT Knowledge Sharing Hub (ACT GenBUDGET, n.d.). It is designed for targeted projects within organisations (e.g., financial allocation models, evaluation of academic work and distribution of research grants) (Steinþórsdóttir et al., 2016), and has been valuable in this context. It was also useful to have an expert on gender budgeting from the City of Reykjavík invited to participate in the second meeting the CoP had in November 2019. She shared her experience of facilitating factors and barriers of the gender impact assessment at the City of Reykjavík, that has been in the process of implementing gender budgeting since 2010. Moreover, additional research funding from the University of Iceland made it possible to invite the CoP members to participate in the complementary research project, Gendersense, and receive additional support with their TIPs. In those cases, a seed partner researcher took a more active role in developing a research design, collecting data, performing the gender impact assessment, and developing objectives and measures.

For the gender impact assessment of the TIPs, CoP members often needed access to information and data on the distribution of financial funds and other resources, such as time, space, equipment, and activities. No less important is the information on the policies and the decision-making that guide that distribution. However, the GenBUDGET members met difficulties in accessing data and information, an obstacle that intensified in the COVID-19 pandemic (e.g., Carmichael and Taylor, 2020). Some CoP members experienced indifference and passivity, as this type of data and information is often perceived as sensitive or not something of importance to progress gender equality. Being a part of a cross-national EU funded project as ACT, provided the GenBUDGET CoP with an underpinning which seems to push forward this work. This has been used by the CoP to put pressure on the administration as the support from the rector or the chancellor gives legitimacy. The importance of gatekeepers has been emphasised in the CoP meetings, thus, to create contact with people in different positions. Through the TIPs work, the University of Iceland's gender equality officer realised his important role as a gate-keeper and that he contributes 'most by getting information and arranging connections because of my positions within the administration' (Guðmundsson, 2021). Working on the TIPs has also proven to be a good tool to increase awareness of key stakeholders and push forward the gender equality agenda within the research organisations. As described by the University of Southern Denmark's CoP Member:

> In the process however, I experience an ability to massage the gender equality (GE) agenda into the organisation organically, where I am able e.g., to catch misconceptions, ensure immediate relevance to individual stakeholders, or explain aspects that require more time than might be available in larger workshops. […] This has increased my awareness of how gender equality work sometimes creates a ripple effect, where

interviewees who I initially perceived to be 'information providers' realize that they themselves, or their colleagues, might benefit from gender equality measures, they share the message with colleagues, qualify my methods, or affect the process in other ways.

(Bjelskou, 2021)

The TIPs have been successful in challenging supposed gender-neutral practices (O'Hagan, 2018) and revealing inequality regimes (Acker, 2006). An example is a TIP at University of Birmingham where the workload allocations system is presented as a tool to enable transparency and fairness. However, the TIP outcomes show that the model design and its implementation in a gendered organisational setting result in gendered outcomes benefiting men and those that structure their academic working lives around masculine norms. This 'inequitable modelling' corresponds with the experiences of academics, especially women, who see the workload allocation models as opaque and unfair (Steinþórsdóttir, Carmichael, and Taylor, 2021). Another example are the TIP findings from the CoP members Carlos III University in Madrid, showing gender different outcomes of the pay mechanisms (Alameda and Pérez del Prado, 2020).

The GenBUDGET public webinars have contributed to the development of shared resources and played a key role in the consensus building of gender budgeting practices within the CoP. There the CoP members had an opportunity to present their findings and experiences of working on the TIPs and receive feed-back on their findings, the measures they are developing and the organisational response to the project. Moreover, the CoP members have been active in sharing information about their work within their institutions and to the broader society. Since September 2020, GenBUDGET members have shared the responsibility for the CoP's blog posts on the Knowledge Sharing Hub. The dissemination has increased awareness within the wider academic community of the gendered outcomes and gender biases of certain decision-making progresses. By presenting the findings within the CoP, their organisations and with the wider academic community the CoP members are increasing the visibility and lowering the legitimacy of the financial managerial mechanisms. These are factors that are essential to change the inequality regimes (Acker, 2006).

Some of the CoP members have mentioned that it is vital to have the TIPs, as they are helping them to gain leverage and get things done within their organisations. Drawing on the gender impact assessment of their TIPs, the CoP members formulate measures that have the potential to enhance gender equal outcomes that must be implemented by the governance of the research organisations. However, the implementation is not always as straightforward as it sounds as the measures to enable structural change are often in conflict with the organisational agenda. Such as in the case of the TIP at the University of Iceland, that found that women have poorer employment and recruitment opportunities than men after PhD graduation. Solutions to the gendered outcomes of the PhD programmes may not be in line with the University's policy objective of increasing the number of PhD graduates (Steinþórsdóttir and Guðmundsson, 2021). Therefore there is a risk of an 'implementation gap', which refers to the fact that the existence of good gender equality policies, does not necessarily guarantee the implementation of such policies in everyday situations (Brorsen Smidt, Pétursdóttir and Einarsdóttir, 2017).

There are more hindrances of various kinds. Although most people speak very positively of equality and diversity in research organisations and are interested in gender budgeting, some of the CoP members have discussed how they face challenges and difficulties on many levels in their work. One example is small units with few employees which do not have the resources for one extra person and face resistance from decision-makers that have the authority to block gender equality implementation processes. Therefore, to get around the organisational barriers, some CoP members need to find ways to convince top decision-makers to do gender budgeting. Nevertheless, the CoP members are creating a basis for action, communication, problem solving, performance, and accountability (Wenger et al., 2002). While it may take time and a lot of small steps to reach the implementation stage, it is far from unsurmountable, according to the 'small wins' model of change (Correll, 2017).

While the GenBUDGET members are positive towards the TIPS, the diversity of projects is also a weakness as the CoP members sometimes find it difficult to engage in discussions. This means that although the TIPs have been successful in a number of ways there is clearly room for improvement in formation and the implementation.

Conclusion

This chapter has sought to provide an insight into the potential for an international CoP to harness inter-institutional cooperation and adopt gender budgeting in research organisations. It has shown how the CoP approach has created opportunities in developing gender budgeting practices to challenge gender biases in research organisations. Gender budgeting is a strategy to get policymakers to consider, reveal, and rethink ingrained gender biases in decision-making processes. The gendered experiences and lived realities of people in research organisations prompts policymakers to engage with gender issues. Gender budgeting offers an open, democratic, and transformative process (O'Hagan, 2018) that has just begun its inroad into research organisations. Furthermore, TIPs, which are specific for the GenBUDGET CoP, created a dynamic and suitable framework and basis for a) the GenBUDGET's community building and b) the development and consensus building of gender budgeting practices in research organisations. This in line with Wenger et al. (2002) argument, that successful practice coincides with community building.

The TIPs demarcated GenBUDGET's activities and formed the basis of the CoPs interactions, activities and thinking, which is important to create a common foundation and to allow members to work together effectively (Wenger et al., 2002). As the CoP members had different knowledge and experience about gender budgeting at the beginning of the CoP's work, the process of the TIPs was unique and individual for each participating member. The GenBUDGET members shared insights of their TIP's progress, such as the data collection, methods, findings, supporting factors and institutional hindrances, enabling fruitful and supportive discussions within the group. The TIPs allowed for differences in competence and experiences, giving all CoP members an opportunity to contribute to GenBUDGET. At the same time as the GenBUDGET members are positive towards the TIPS, the diversity of projects may also be a weakness as it can be difficult to deepen the insights and engage in more specified discussions. The ACT support has also contributed to the development of GenBUDGET. ACT provided a venue for the CoP to meet on regular bases and organise its activities, to plan and hold monthly meetings, workshops, and webinars. Moreover, despite the pandemic, which has influenced the GenBUDGET's development extensively, GenBUDGET has been able to continue its activities, although remotely, and the CoP is growing.

With the TIPs, the GenBUDGET members have got hands on experience of gender budgeting work, insights and discussion that have resulted in development and consensus building of gender budgeting practices in research organisations. Although GenBUDGET has not yet reached the implementation stage of the TIPs, the CoP is creating and sharing knowledge, and thus, creating conditions for continued development and progress. This is reflected in the CoP members' various achievements, such as presentation in GenBUDGET webinars and writing of articles. By this, the CoP members contribute to the visibility of gender equality issues and gender biases in decision-making in their organisations. This is an important step in decreasing the legitimacy of those inequalities (Acker, 2006). As the stimulation of learning processes and knowledge sharing activates structural change (Barnard, Hassan, Dainty, Polo, and Arrizabalaga, 2016), the GenBUDGET CoP has the potential to increase competence in the design, implementation, monitoring, and evaluation of gender budgeting for institutional change (Palmén et al., 2019). By this, the GenBUDGET CoP has contributed to progressing the work towards more gender equal decision-making in research organisations.

Note

1. The MoU is a formal documented understanding of mutual responsibilities, obligations, and benefits between a representative (e.g., legal representative, Rector, Dean, etc.) from all the participating organisations, and representative from the ACT Project and the seed partner (University of Iceland).

References

Acker, Joan. (2006). Inequality regimes gender, class, and race in organizations. *Gender & Society, 20*(4), 441–464.

ACT GenBUDGET. (n.d.). GenBUDGET: Gender budgeting in research organisations. Retrieved from: https://genbudget.act-on-gender.eu/

ACT. (2019). 1st ACT International Synergy Conference. Retrieved from: https://www.act-on-gender.eu/act-events/1st-act-international-synergy-conference

Addabbo, Tindara, Valeria Naciti, Guido Noto, & Carlo Vermiglio. (2020). Budgeting for gender equality in Research Performing Organizations. *Politica Economica, 36*(3), 417–438.

Addabbo, Tindara, Paula Rodríguez-Modroño, & Lina Gálvez-Muñoz. (2015). Gender budgeting in education from a wellbeing approach: An application to Italy and Spain. *Politica Economica, 31*(2), 195–212.

Alameda, María Teresa, & Daniel Pérez del Prado. (2020). Gender pay gap and gender budgeting: a look at the University's data [blog]. Retrieved from: https://genbudget.act-on-gender.eu/Blog/gender-pay-gap-and-gender-budgeting-look-universitys-data

Barnard, Sarah, Tarek M. Hassan, Andrew R. J. Dainty, Lucia Polo, & Ezekiela Arrizabalaga. (2016). Using communities of practice to support the implementation of gender equality plans: Lessons from cross-national action research project. Presentation at the GenderTime International Conference 2016. Paris, France, 29–30 September 2016.

Bjelskou, Peter. (2021). Relational gender equality work [blog]. Retrieved from: https://genbudget.act-on-gender.eu/Blog/relational-gender-equality-work

Brorsen Smidt, Thomas, Gyða Margrét Pétursdóttir & Þorgerður Einarsdóttir. 2017. "How do you take time? Work–life balance policies versus neoliberal, social and cultural incentive mechanisms in Icelandic higher education." *European Educational Research Journal, 16*(2–3): 123–140. https://journals.sagepub.com/doi/pdf/10.1177/1474904116673075

Bryman, Alan. (2012). *Social Research Methods* (4th edition). Oxford: Oxford University Press.

Budlender, Debbie, & Guy Hewitt. (2002). *Gender Budgets Make More Cents.* London: Commonwealth Secretariat.

Carmichael, Fiona, & Scott Taylor. (2020). COVID-19, ways of working, and gender equality in research organizations [blog]. Retrieved from: https://genbudget.act-on-gender.eu/Blog/covid-19-ways-working-and-gender-equality-research-organizations

Correll, S. J. (2017). SWS 2016 Feminist Lecture: Reducing gender biases in modern workplaces: A small wins approach to organizational change. *Gender & Society, 31*(6), 725–750.

Council of Europe. (2010). *Recommendation 1921.* Strasbourg: Council of Europe.

Erbe, Birgit. (2011). Gender budgeting as a management strategy for gender equality at universities – with the examples of Austria, Germany and Poland. In Hildegard Macha, Claudia Fahrenwald, & Quirin J. Bauer (Eds.), *Proceedings of Gender and Education: Towards New Strategies of Leadership and Power. Second International Conference of the Women Leading Education (WLE) Network* (pp. 57–62). Berlin: Druck und Verlag: epubli GmbH.

Erbe, Birgit. (2015). Gender mainstreaming in public financing of universities: Central findings for Germany. *Politica economica*, *31*(2), 213–232.

GARCIA. (n.d.). GARCIA project. Retrieved from: http://garciaproject.eu

Guðmundsson, Sveinn. (2021, April 3rd). Getting comfortable with feeling uncomfortable [blog]. Retrieved from: https://genbudget.act-on-gender.eu/Blog/getting-comfortable-feeling-uncomfortable

Heijstra, T. M., Þorgerður Einarsdóttir, Gyða Margret Pétursdóttir, & Finnborg Salome Steinþórsdóttir. (2017). Testing the concept of academic housework in a European setting: Part of academic career-making or gendered barrier to the top? *European Educational Research Journal*, *16*(2–3), 200–214.

Heijstra, Thamar Melanie, Finnborg Salome Steinþórsdóttir, & Þorgerður Einarsdóttir. (2017). Academic career making and the double-edged role of academic housework. *Gender and Education*, *29*(6), 764–780.

Lave, Jean, & Wenger, Etienne. (1991). *Situated learning: Legitimate Peripheral Participation*. Cambridge: Cambridge University Press.

O'Hagan, Angela. (2018). Conceptual and institutional origins of gender budgeting. In: Angela O'Hagan and Elisabeth Klatzer (Eds). *Gender Budgeting in Europe: Developments and Challenges* (pp. 19–42). New York, London: Palgrave Macmillan.

O'Hagan, Angela. (2021, March 1st). Transformative potential of gender budgeting in HEIs [blog]. Retrieved from: https://genbudget.act-on-gender.eu/Blog/transformative-potential-gender-budgeting-heis

Palmén, Rachel, Caprile, Maria, Panadès, Rosa, Riesco, Julia, Pollitzer, Elizabeth, & Vinkenburg, Claartje J. (2019). ACT conceptual framework – Gender equality and Communities of Practice. *Zenodo*. https://doi.org/10.5281/zenodo.3235296

Rothe, Andrea, Birgit Erbe, Werner Fröhlich, Elisabeth Klatzer, Zofia Lapniewska, Monika Mayrhofer, Michaela Neumayr, Michaela Pichlbauer, Malgorzata Tarasiewicz, Johanna Zebisch & Maciej Debski. (2008). *Gender Budgeting as a Management Strategy for Gender Equality at Universities: Concluding Project Report*. Munich: Frauenakademie München e.V. München

Sawer, Marin, & Miranda Stewart. (2020). Gender budgeting. In Marin Sawer, Fiona Jenkins, and Karen Downing (Eds.), *How Gender Can Transform the Social Sciences: Innovation and Impact* (pp. 117–126). Cham: Palgrave Pivot.

Schreier, M. (2013). *Qualitative Content Analysis in Practice*. London: Sage.

Steinþórsdóttir, Finnborg Salome, Thomas Brorsen Smidt, Gyða Margrét Pétursdóttir, Þorgerður Einarsdóttir, & Nicky Le Feuvre. (2018). New managerialism in the academy: Gender bias and precarity. *Gender, Work & Organization*, *26*(2), 124–139. doi:10.1111/gwao.12286

Steinþórsdóttir, Finnborg Salome, Fiona Carmichael, & Scott Taylor. (2021). Gendered workload allocation in universities: A feminist analysis of practices and possibilities in a European University. *Gender, Work & Organization*, 1–17. doi:10.1111/gwao.12709.

Steinþórsdóttir, Finnborg Salome, Þorgerður Einarsdóttir, Thamar Melanie Heijstra, Gyða Margrét Pétursdóttir, & Thomas Brorsen Smidt. (2019). Gender budgeting to expose inequalities in a precarious academia – and redistribute resources to effect change. In Annalisa Murgia & Barbara Poggio (Eds.). *Gender and Precarious Research Careers: A Comparative Analysis* (pp. 83–110). Milton Park, New York: Routledge. Taylor & Francis Group.

Steinþórsdóttir, Finnborg Salome, & Sveinn Guðmundsson. (2021). Gender budgeting in the Unviersity of Iceland: Gendered career progression of PhD holders. Presentation at the ACT GenBUDGET webinar, 23 February 2021.

Steinþórsdóttir, Finnborg Salome, Thamar Melanie Heijstra, & Þorgerður Einarsdóttir. (2017). The making of the 'excellent' university: A drawback for gender equality. *Ephemera, Theory & Politics in Organization, 17*(3), 557–582.

Steinþórsdóttir, Finnborg Salome, Thamar Hejstra, Þorgerður Einarsdóttir & Gyða Margrét Pétursdóttir. (2016). Gender budgeting in academia – Toolkit. Garcia working papers 14. http://garciaproject.eu/wp-content/uploads/2016/12/GARCIA_D5.3-Gender-budgeting-in-academia-toolkit.pdf

Stotsky, Janet Gale. (2016). Gender budgeting: Fiscal context and current outcomes. In *IMF Working Paper WP/16/149*. Washington: International Monetary Fund.

Thomson, Aleksandra, & Rabsch, Kathrin (2021). *ACT – Community of Practice Co-creation Toolkit*. Zenodo. https://doi.org/10.5281/zenodo.5342489

Wenger, Etienne. (2000). Communities of practice and social learning systems. *Organization, 7*(2), 225–246. DOI: 10.1177/135050840072002

Wenger, Etienne. (2001). Supporting communities of practice: A survey of community-oriented technologies. Report to the Council of CIOs of the US Federal Government.

Wenger, Etienne, Richard McDermott, & William M. Snyder. (2002). *Cultivating Communities of Practice: A Guide to Managing Knowledge*. Boston: Harvard Business School Press.

Wenger-Trayner, Etienne, & Beverly Wenger-Trayner. (2015). Communities of practice: A brief introduction. Retrieved from: https://wenger-trayner.com/wp-content/uploads/2015/04/07-Brief-introduction-to-communities-of-practice.pdf

9 Sustainability and gender equality

A co-creation and Communities of Practice approach

Areti Damala, Chloé Mour
and Anne-Sophie Godfroy

Introduction

Strategies for Sustainable Gender Equality (STRATEGIES) is one of the eight Communities of Practice (CoPs) supported by the ACT project, funded under the Horizon 2020 framework programme (see the introduction for an explanation of the project). STRATEGIES placed particular emphasis on the question of 'gender equality' and 'sustainability' in terms of project management. The ACT seed partner responsible for the coordination of the CoP is the French National Centre for Scientific Research (Centre National de la Recherche Scientifique) and the Republic of Knowledge research team – laboratory from the Ecole Normale Supérieure in Paris, France. The lab gathers philosophers, humanists and researchers on mathematics, biology, and computer science who share a common interest in epistemology as well as in history of science and technology. The dissemination of scientific knowledge provides another convergence point for the research carried out in the lab and within this scope the broader contextual, institutional and organisational learning around gender equality and Gender Equality Plans (GEPs) is of particular interest in the host research institution.

Much like all the other seven sister CoPs, STRATEGIES inception was founded upon a CoP approach. A CoP is 'a group of people who share a concern, a set of problems or a passion about a topic, and who deepen their knowledge and expertise by interacting on an ongoing basis' (Wenger, 2000). CoPs are often characterised by a 'shared identity,' as well as a collective intention around a problem. The introduction chapter provides a more in-depth presentation of the way the CoP theory shaped the goals and the ambitions of the ACT project and its supported CoPs. The shared domain for STRATEGIES is promoting gender equality from a sustainability perspective. The collective intention was to identify, share and leverage strategies for sustainable gender equality. The question of continuities and discontinuations in gender equality projects, initiatives and plans formed the core of the shared identity. The goal and hope were that by bringing together gender equality

DOI: 10.4324/9781003225546-9

practitioners from around Europe with an intense interest and preoccupation with sustainability, gender equality and GEP, best practices and lessons learned would be shared, inspiring new approaches that can guarantee sustainability in GEPs and policies in higher education (HE) research and innovation (R&I).

The text revisiting the life of STRATEGIES uses a reflective writing approach focusing on empirical and experiential reflections, ideas and findings from coordinating and facilitating the CoP: all three authors have acted as CoP facilitators and worked closely to enable knowledge exchange, events and activities of the life of the CoP. Reflective notes and a reflective diary were kept throughout the life cycle of the project. Reflective diaries are an interesting tool for facilitating and assessing reflection (Tang, 2020) as well as for assessing what has been learned (Wallin & Adawi, 2018). Particularly useful were also the blog entries of key moments in the life of the CoP available through the main ACT project website. Another resource we used was the ACT project evaluation reports (also available from the project website), which helped us gain an additional perspective through the eyes of ACT colleagues not directly involved in the life of the CoP. Within the framework of the ACT project, brainstorming and reflective, collaborative writing among different ACT CoP facilitators also laid the foundations for the writing of the ACT Policy Brief 'How to support CoP for driving institutional change towards gender equality' which was prepared by STRATEGIES in collaboration with Alt+G and LifeSciCop sister ACT CoPs (Mihajlović Trbovc et al., 2021).

Strategies for sustainable gender equality: Setting up the community of practice

STRATEGIES set itself the goal of addressing the complex, multifaceted topic of sustainability. This is reflected in the chosen name of the CoP 'Strategies for Sustainable Gender Equality.' The main angle from which we wanted to approach the topic concerns the knowledge, know-how, skills, competencies, policies and resources generated throughout various projects during their full life cycle, despite and beyond their end. A common example is this EU funded projects, however similar tendencies can be observed with projects funded by national bodies and authorities. What happens when projects end? What can we do to assure continuity? How can we avoid reinventing the wheel by making great use of available resources, tools and outcomes? Sharing experiences of continuity and discontinuity helps members of the CoP to better understand reasons for discontinuity and to share strategies or contingency plans to avoid this discontinuity.

In light of the above, the name we chose for our CoP suggests the three aims:

- 'Strategies' recall the fact that the strategy for gender equality has to be part of the overall strategy of the institution. If gender equality is considered as a side policy or an outsider strategy, carried out by outsiders, there is a high probability the policy will disappear when the project ends and/or when the key person leaves. We wanted to share ideas to make gender equality part of the overall long-term strategy of our institutions.

- 'Sustainable equality' is an invitation to promote sustainable policies and to institutionalise gender equality in the day-to-day management of HE, R&I environments. An institutionalised policy is less subject to personnel turn-over or political change, and benefits of secure funding and staff. We wanted to share paths to transform gender policies from optional policies into institutionalised policies.
- 'Sustainable equality' also means cultural change towards gender equality, evaluation and the monitoring of GEPs. Evaluation, monitoring and impact assessment need to address bottom-up and top-down cultural change and address issues of how they can reinforce each other. It may be grass root initiatives of women scientists who promote a workshop to work on gender or students' initiatives to create a new research seminar. It may be also executive level initiatives to promote new procedures in evaluating or recruiting, or considering the gender dimension in research as much as in academic teaching. Evaluation also needs to analyse how the different scales (micro, meso and macro) interact to produce sustainable progress and change.

Strategies for sustainable gender equality: Community

STRATEGIES was built around the former consortium of the GenderTime project (2013–2016) while new members were met and approached at a workshop, at the ACT first Synergy Conference in Brussels on 25 February 2019. The basis of these preliminary exchanges was provided by the notion of sustainability related to gender equality, and sustainability as outlined earlier. The main shared common goal and vision was to reflect on strategies ensuring sustainable gender equality policies, with a particular emphasis on HE and R&I environments. We were interested in figuring out how to improve gender equality for various stakeholders opening to research and academic staff, administrative staff and students. We also wanted to encourage a cross-border, cross-disciplinary collaboration which would encourage Strategies members to reflect, grounded on their own experiences, learning and know-how, what sustainability means for academic gender equality projects and initiatives and how a sustainable change towards a culture of gender equality may be achieved, developed and measured.

By the last months (October 2021) of the ACT project, STRATEGIES counted 14 members from France, Sweden, Serbia, Cyprus, Germany, Italy, Belgium and the United Kingdom, while a close collaboration with Eurodoc (the European Council for Doctoral Candidates and Junior Researchers) offers our CoP a fresh, Early Career Researcher and Investigator perspective.

It should be noted that among these 14 members, 6 members (including the coordinator) come from France and had already collaborated and worked together both in the context of other gender equality EU funded projects such as INTEGER, GenderTime, TRIGGER or EGERA. Sustained participation in both national and EU projects and initiatives created the will to work and join forces around gender equality and GEP while addressing the challenge of organisational settings and academic cultures characterised by modest resources and centralised management, as well as discontinuity in implementing GEP and related assessment and monitoring.

The 14 CoP members (including our own institution) had various profiles including universities (Deusto, Université de Strasbourg, Université Paris-Est Créteil, Oxford Brookes University, Universita degli Studi di Padova, Örebro University, Wuppertal University and Frederick University), associations (Femmes et Mathématiques, Femmes et Sciences), networks (CPED, Eurodoc) and technological and research organisations (Mihalo Pupin Institute). The countries represented in the consortium were: Sweden, France, Italy, Spain, the United Kingdom, Germany, Cyprus and Belgium (for Eurodoc) with France counting five members.

A very important feature and element we believe sets our CoP apart, is the number of associations and networks we host (n = 5), acting themselves as CoPs: 1. Eurodoc, 2. CPED (Standing Conference for Equality and Diversity), 3. Femmes & Sciences (Women and Science), 4. Femmes & Mathématiques (Women and Mathematics). A fifth network is the French CNRS (French National Centre for Scientific Research) which also constitutes a network of labs, researchers and research units in France. STRATEGIES proudly has as a member Eurodoc, the European Council for Doctoral Candidates and Junior Researchers, a federation of 28 national associations of doctoral candidates and early career researchers (pre-doctoral or post-doctoral researchers employed on a temporary basis) of the European Union and the Council of Europe. This feature of STRATEGIES as a network of networks gave the potential and opportunity to reach out to multiple institutions in countries currently not present in the CoP. It gave us the opportunity to disseminate news, surveys, best practices, guidelines and practices with an extended network that reaches researchers and practitioners through various stages in their careers. It also allowed us to examine recurring challenges and themes under a regional, national and cross-border perspective.

The mix of networks and more traditional institutional members allowed both the identification of very specific case studies and the possibility to disseminate recommendations on a large scale. The interest of members and their reasons for participating stemmed from their desire to learn from others and share experiences, and also to disseminate their own good practices and recommendations. The participation of the four networks also allows an immediate dissemination and scaling up of activities and initiatives judged interesting and pertinent for the network members.

The domain: Sustainability and gender equality

Sustainability and project management

Seconding and complementing the CoPs approach (Wenger, 2011), the concept of sustainability was paramount to the identity and aspirations of all STRATEGIES members. Although the notion of sustainability can vary in terms of context, situation, scale, time and space, sustainable development in project management can be defined as a 'continuity of economic, social, institutional and environmental aspects of the human society and the non-human environment in such a way that the needs of the present are met without compromising the needs of the future' (Morfaw, 2014). In this sense, sustainability can be considered as a new form of humanism putting forward the notions of prosperity, peace, equality and well-being for all: individuals, communities, societies and institutions alike.

Gender equality and Sustainable Development Goals

Nowhere is this more apparent than in the 17 Sustainable Development Goals adopted by the United Nations in 2015. Gender equality is recognised as a key challenge (Goal 5: Gender Equality) crosscutting other goals and priorities. The Sustainable Development Agenda emphasises that 'realizing gender equality and the empowerment of women and girls will make a crucial contribution to progress across all the Goals and targets' and stresses out that work shall be carried out for 'a significant increase in investments to close the gender gap and strengthen support for institutions in relation to gender equality and the empowerment of women at the global, regional and national levels.' A prevision for the systematic mainstreaming of a gender perspective in the overall implementation of the agenda is also discussed, setting as a goal the elimination of any kind of gender-based discrimination. More broadly speaking it is also recognised that achieving gender equality will help progress across all sustainable development goals and targets (Blewitt, 2018). Of particular importance in terms of gender equality and GEPs are also Goals 10: Reduced Inequalities and Intersectionality, 11: Sustainable Cities and Communities, 16: Peace, Justice and Strong Institutions and 17: Partnerships for the Goals (Esquivel & Sweetman, 2016).

Sustainability in the previous experiences of STRATEGIES members

As already highlighted, almost all members of STRATEGIES had previous experience and involvement in gender equality projects, plans and initiatives at a national and/or European level. Common in these experiences was a larger preoccupation with regards to gender equality and project sustainability. More broadly speaking and within the project management literature, project sustainability is examined and discussed as an 'integrated process involving social, economic, cultural, legal, political, health, environmental, financial, and a host of other factors which can facilitate continuity and sustainability of an organisation, system, structure, or institution' (Morfaw, 2014).

This is particularly true for GEP and gender equality projects: STRATEGIES came as an attempt to provide some answers around a common concern of the CoP members from previous structural change projects and the question of the sustainability of gender equality policies: most STRATEGIES members had experienced a huge boost towards gender equality during an EU funded project followed by a clear slowdown after the end of the project due to the finalising of human and financial resources provided by the project funding. STRATEGIES was founded on the basis of the conviction that even if this outcome is not easy to manage, it is quite foreseeable while preventive measures, actions and initiatives can be imagined to deal with it, such as managing to anticipate the end of the project and obtain from the management of the institution the resources to continue the work, to which the European funding theoretically committed. Other contingencies are less predictable and can undermine gender policies, as changes in political teams, either at the institutional level (e.g., in France, university presidents are elected every 4 years), or at the national level, when anti-gender coalitions come to power. Another common contingency is the turnover of key persons to implement gender policies, and the subsequent lack of follow-up.

This structural and organisational challenge has been recently clearly recognised by the European Commission: 'Reviews of GEP projects have shown that the long-term impact of the structural changes initiated cannot be judged right after the end of a project, and that devising effective strategies to ensure the sustainability and institutionalisation of the GEPs are crucial to achieve lasting transformation.' (European Commission, 2020). The document 'Gender Equality, Achievements in Horizon 2020 and recommendations on the way forward' states that several projects report 'a major obstacle to institutional change due to a change in the top management of the implementing institution during the project. This can be a crucial issue if the newly appointed management does not consider gender equality as a priority. To mitigate this risk, the most effective long-term strategy is the early push for an institutional gender equality policy with defined and binding commitments and targets. An integrated, institutionalised, policy in management structures, and a GEP that is made publicly available, making successors likewise accountable for gender equality matters.'

Strategies for Sustainable Gender Equality as a CoP was formed around recurring challenges with regards to continuities, discontinuities and disruptions of gender equality projects and plans. What strategies can be developed to ensure continuity in gender equality policies? How can these consistently remain on the agenda for HE management at the regional, national and European level? Can we prevent new management teams from undoing what has been done? What strategies can be developed to mitigate the consequences of such contingencies and ensure continuity?

Sustainability is not just a concern for GEP or gender equality projects. In fact, the 'traditional' notion of project sustainability so far is mostly related with what happens once a project or initiative is over. In too many projects, when the funding is over, or when the top management is replaced (for example because of the election of a new university president or dean), or when key people move to another place, or when the overall political context moves to another top priority, the project declines and the actions are not followed up. This echoes the three recommendations provided by Silvius and Schipper (2014) who argue that considering sustainability in project management implies: i. 'a shift of scope in the management of projects: from managing time, budget and quality, to managing social, environmental, and economic impact.' ii. 'a shift of paradigm of project management: from an approach that can be characterised by predictability and controllability, to an approach that is characterised by flexibility, complexity and opportunity.' iii. 'a mind shift for the project manager: from delivering requested results, to taking responsibility for sustainable development in organisations and society.'

An additional concern common in our CoP was that many members had the feeling that the recommendations which existed at the time when the community was formed, were not always applicable in their own institutional context. For example, policies designed and implemented in countries in which universities are self-regulated, autonomous, self-financed and self-managed are probably not applicable the same way in HE environments which are regulated at a national level, with limited resources allocated. In such contexts, lobbying at parliamentary or governmental levels is essential to obtain gender equality inscribed in the laws and the national regulations.

Sustainable gender equality within the European research area

The fact that the sustainable project management concern experienced from STRATEGIES members is very timely is also reflected in several EU regulatory texts. The European Commission recently reaffirmed the need for both 'sustainable cultural and institutional change' (European Commission, 2020). Despite the fact that the implementation of GEP in research performing organisations (RPOs), including universities and research funding organisations (RFOs) was already supported in FP 7 (the European 7th Framework program, it lasted until 2013), new, additional previsions and mechanisms were put forward in Horizon 2020 in order to support institutional changes and address gender inequalities at all levels: a network of national representatives and resource centres on gender were established, potential gender biases in the allocation of grants were scrutinised, while training and capacity building in R&I was supported (European Commission, 2020).

Although much has been achieved for gender equality in R&I under the FP7 and Horizon 2020 programmes, inequalities still persist (see introduction to this volume). Strengthened provisions were decided within the new Horizon Europe programme, as a wider scale implementation was judged essential for reducing inequalities, among which some for the very first time: the most notable is the fact that, starting in 2022, public bodies, research organisations and higher education establishments will be required to have a GEP in place in order to ensure sustainable institutional change (European Commission, 2021). This sets a new basis and paradigm for gender equality in Europe with important consequences towards achieving sustainable gender equality as well as pertinent, related and sustainable institutional change. Furthermore, all provisions for gender equality in research and innovation under Horizon Europe will contribute directly to the United Nations' Sustainable Development Goal (SDG) 5 on Gender Equality and Women's Empowerment, and to all SDGs, as gender equality is a necessary foundation across SDGs (European Commission, 2020).

National initiatives and gender equality strategies are also headed towards the obligatory implementation of GEPs: for example, universities in France were invited to design and implement a GEP beginning from 2021. Ensuring that GEPs are in place will provide a new basis for discussing sustainability in ensuring gender equality and fighting discriminations. We believe that STRATEGIES has been a precursor of this movement towards achieving greater and more sustainable gender equality for all.

The practice: Leveraging sustainability for gender equality with co-creation

Capital to the life of the community was a co-creation approach nourished by the co-creation materials and methods that were provided by the core ACT project partners, particularly those who led the co-creation process (see Chapter 4 in this volume). Practice in STRATEGIES was also nurtured by the various backgrounds and disciplines practiced by the participants: philosophers, historians of history and science and technology, literature, Science-Technology-Engineering and Mathematics (STEM) researchers, humanists, cultural communication specialists, sociologists, gender experts) with various degrees of expertise and experience from day-to-day practice in carrying out research, academic teaching and mentoring, administrative and governance responsibilities, in various environments (research, innovation, higher education, policy making) and cultural contexts. This gave us exciting challenges we were eventually able to cope with. Transnational, multi-institutional and interdisciplinary CoPs may face such challenges (Thomson et al., 2021).

The community functions mainly in English, although French is commonly privileged for communicating with all French partners. Within the measure of possible CoPs, members were encouraged to bring into the life of the CoP their own experiences, lessons learned, needs, weaknesses, projects, initiatives, know-how while guests and invited speakers were often invited to present, teach and reflect on lessons learned highlighting challenges, gaps and potential steps forward. For this to occur, we tried to create a safe-as-much-as-flexible space and cater to each participant needs. Monthly meetings and get togethers were complemented by one-to-one meetings as needed.

Within the CoP, get togethers between members on topics of common interest identified by a co-creation, bottom-up approach. These meetings brought into the picture conceptualisations, theoretical, methodological and empirical approaches featured in flagship European and national projects and initiatives. In terms of who is representing the institutions mentioned above, the policies differ: some institutions are represented in all meetings and communications by the same representative, while others are represented by more than one member depending on the occasion or the topic treated.

For setting-up the CoP agenda and work-plan, we opted for a co-creation, bottom-up approach (Sanders and Simons, 2009), largely inspired by the spirit and techniques offered at the ACT Co-Creation toolkit (see Chapter 4).

We used the first/kick-off physical meeting in December 2019 for brainstorming, then co-constructing the main topics upon which we aimed to scaffold the CoP efforts and energy, among a variety of topics brought to the meeting by CoP members.

Two axes were identified by STRATEGIES members during the first, kick-off meeting: An Evaluation and Impact Assessment and B. Moving towards Structural Change regarding Training, Recruitment and Career Progression. Institutional and organisational learning was identified as a third axis, cross-cutting the two main topics identified.

Organisational learning has been defined as a continuous process of detecting and correcting errors (Argyris and Donald, 1978). According to the early organisational learning theory, open communication can hinder progress based on denial of real problems and denial for examining one's own contribution to a problem. More recent theories approach the question of institutional and organisational learning more as a process gained throughout the full life-cycle of a project rather than as a final, definitive outcome (Damala et al., 2021). With these in mind, we encouraged the CoP members who participated in the first, kick-off meeting to reflect on their own practices and challenges faced in implementing GEPs. Here are some of the core questions that emerged during the workshops which were organised on the occasion of the kick-off meeting: How can we define the notion of change (including institutional and organisational learning)? What constitutes a 'strategy'? How can we train and educate research, teaching, administrative staff on gender equality? What are the pitfalls of the current situation today in Europe and overseas? How do we move towards change integrating a micro, meso and macro level from grassroot initiatives to large scale national and EU policies? How can we avoid discontinuities and better institutionalise gender equality policies?

To this cross-cutting topic, other questions were associated, pertinent with the two main axes:

a Evaluation and Impact Assessment: How do we measure and evaluate gender equality practices in HE and R&I environments? How do we leverage and demonstrate the impact of the policies, measures and strategies adopted at a national and European or cross-border level? Should we share common indicators? The diversity of indicators is both a chance to get tailor-made tools and a challenge to build an overall picture of the situation. Experiences with existing awards and labels, existing monitoring and measurement tools were judged as important topics to share in the CoP. Other questions brought forward by the CoP members were the transferability of evaluation and impact assessment practices and indicators from one context to another and methodological issues such as the quality of data and collection and the challenges regarding privacy, confidentiality, the survey fatigue leading to poor answering rate, etc. Alternative strategies as participatory audits or formative evaluations were also flagged as important to reflect upon.

b Moving towards structural change regarding training, recruitment and career progression: As already mentioned, Eurodoc was one of the CoP members that was welcomed in STRATEGIES. Other members, particularly those representing HE institutions were also aware of the challenges faced by Early Career Investigators and Researchers. An important group of open questions were identified in the kick-off meeting such as: What can be done to assist researchers throughout all stages of their careers (including early career researchers)? What are the current impediments and how can they be proved and demonstrated? What can be done to create mechanisms and conditions that guarantee hurdle-free progression with one's career at all levels and stages? The question also covers issues around the gender dimension of work-life balance. In line with the EU policies, beyond supporting and mentoring persons, beyond cultural change regarding gender regimes, our aim is to change the institutions and to stimulate a structural change. How can we change the institution if the institution produces discrimination? These challenges are especially important for Eurodoc and the early career researchers represented, also identified in the relevant literature (Bozzon et al., 2016; Murgia and Poggio, 2019).

With these in mind, we collaboratively decided to host the next meetings around these two topics, always under the lens of sustainability in gender equality as well as trying to address broader questions in institutional and organisational learning.

Gender equality, evaluation and impact assessment

After the question of Evaluation and Impact Assessment emerged as one of the most important questions, we decided to dedicate the second, consolidation workshop and get together on this topic. The second (and last physical) meeting of the CoP was equally organised in Paris in March 2020, just a few days before the large majority of European countries entered into COVID-19 lockdowns.

In order to cross-pollinate the reflections, collaboration and work of the members of our CoP, we devised a programme composed of presentations on national and European initiatives on the question of Evaluation and Impact Assessment. Three major European initiatives, that had set up important paradigms, were chosen for inclusion in the programme of the day: The ADVANCE UK Athena Swan initiative and the GEAM tool and the SUPERA and EFFORTI EU projects. All three had addressed – under a different perspective – the question of evaluation and impact assessment.

The ATHENA Swan accreditation scheme, which has grown to be recognised further than the United Kingdom for which it was initially conceived was presented by Kevin Guyan from ADVANCE HE, Edinburgh, United Kingdom with a presentation entitled 'Evaluating gender equality – Advance HE research, surveys and accreditation.' The UK 2010 Gender Equality Act was presented as well as the ways through which the UK legislation encourages and motivates UK institutions to work with gender equality. The question of staff retention and the development of the overall student experience was brought forward. The presentation also covered an overview of the UK 2010 Gender Equality Act 'identity' characteristics. Data gathered around identity characteristics such as gender, race, disability, religion, sexual orientation and others can be used to provide a demographic picture of an organisation, to provide insights around people's lives and their real experiences or be used as a research tool for advancing equality, for example in order to evaluate the success/failure of initiatives and establish what works. Such data can also be used for encouraging diversity by establishing a diversity profile (a demographic picture of an organisation, used to identify gaps/absence and benchmark against comparators) and finally for promoting greater inclusion. The presentation to the CoP members also included an overview of the ASSET 2016 study 'Experiences surrounding gender equality in STEM academia and the intersections with ethnicity, sexual orientation, disability and age' (Aldercotte et al., 2017), which gathered rich data on gender equality in HE and Recruitment, Job and career, Perceived gender equality, Caring responsibilities, Training and leadership, Promotion and development.

The work carried out in ASSET fed into the development of the Gender Equality Audit and Monitoring (GEAM) tool which provides an integrated environment for carrying out survey-based gender equality audits in organisations. The availability of the GEAM survey in several languages might – in the long term – assist in cross-border, comparative overviews of how the tool will be used across different cultural and linguistic contexts. This can be a major opportunity for promoting sustainable gender equality (for details on the development of the GEAM tool, see Chapter 3 this volume).

Inspiring work carried out within the European SUPERA project provoked sparkling discussions within STRATEGIES (Forest and Lombardo, 2012). Maxime Forest from the Paris Institute of Political Studies, Paris, France presented the SUPERA EU Project approach to evaluation for gender equality which aimed at a structural understanding of gender inequalities, stereotypes and biases in research and academia as a cross-cutting issue. The main ambition was to propose a holistic set of measures to assist with building gender sensitive career management and workplaces, transform decision-making towards accountability, transparency and inclusiveness and achieve excellence through strengthening the gender dimension in research and knowledge transfer. Of particular interest is the position that one should bring into the picture the broader context (social, cultural, financial, regional, national), and the fact that innovative measures and initiatives should be disseminated and made known by various relevant stakeholders.

Based on the above, the SUPERA approach to evaluation was presented: This brings together a formative evaluation (which aims to reinforce the capacity of GEP actors and relevant stakeholders to design and set in place efficient changes), support for a strategic framework and thinking of the planned actions/interventions as well as tools supporting both the procedure of validation as well as the set-up and launch of relevant actions and activities around evaluation. The process can be supported by the participation of all involved parties and stakeholders, by raising the capacity of all agents of change to identify and understand windows of opportunity available at an institutional level as well as adopting, promoting and supporting proposed measures at an institutional level as a precaution and preventive measure encouraging sustainability.

A third important influence and contribution came from Susanne Buehrer and the EFFORTI project. The EFFORTI EU-funded project has the particularity of explicitly addressing the question of evaluation and impact assessment of gender equality in R&I in terms of research and hands-on practice. Its main goal was to develop an evaluation framework for establishing a link between Responsible R&I and gender equality based on the fact that more tangible evidence is needed for gender equality as a prerequisite for improved Research and Innovation outcomes (e.g., improved societal relevance of R&I, better contribution of R&I to societal challenges, innovations better suited to markets etc). The project proposed a framework for capturing the complexity of interventions and their impacts in complex systems, putting into the picture gender equality interventions and linking these to outputs, outcomes and impact (Palmén et al., 2020). Establishing appropriate frameworks that can guide evaluation and impact assessment was then recognised as a major opportunity by all STRATEGIES members.

Gender inequalities at the early career stage

As already highlighted, the CoP practice was greatly enriched through the active participation of Eurodoc, representatives. The extensive involvement of several of their core members enabled the integration of PhD and Postdoctoral researchers' challenges with a gender perspective. Eurodoc, and the topic of Early Career Researchers and Investigators (ECRIs), progressively gained an important space within our community. This was partly due to the sustained interaction that happened between Eurodoc's members and two CoP facilitators who were themselves at an early career stage within academia. Sharing the same social status (ECRIs) as well as gender (women) certainly fostered this learning partnership.

The lack of – even absence of – emphasis on early career researchers within gender equality initiatives in HE and R&I is another strong factor accounting for this peculiar dynamic within our CoP. Gender equality projects and plans are mostly designed by and for senior researchers. ECRIs, along with administrative and non-administrative staff as well as students, most of the time constitute a blind spot for gender equality policies. Their specific challenges are hardly monitored and addressed through those policies, because tracking them is a complicated task for institutions. Indeed, ECRIs are often funded by research projects or external stakeholders and do not appear in institutional data sets, since they frequently hold short term positions and are very mobile. Their variety of experiences and career paths is, therefore, invisible in policies tackling gender inequalities. However, the unstable working conditions they often face due to the growing casualisation of the academic workforce and the precariousness of the academic life is profoundly gendered and needs to be addressed (Murgia & Poggio, 2019).

Through various meetings and webinars within our CoP, a first goal was to provide a greater understanding of the gender inequalities at the early career stage and the specific challenges of ECRIs, especially in the aftermath of the COVID-19 crisis (Eurodoc, 2020). The challenges discussed, for instance, the issue of gender and academic mobilities in the context of the internationalisation of EU careers by exploring the career norm of geographical 'mobility' and its gendered assumptions and effects (Sautier, 2021). Simultaneously, a second objective was to share experiences and strategies aiming at establishing sustainable gender equality at an early career period. Several CoP webinars provided the opportunity to exchange information on both individual (mentoring programmes) and structural (implication of ECRIs in decision-making processes and science policy) measures implemented by CoP members in their institutions or by other researchers and academic staff participating as guest speakers.

At a more practical – empirical and experiential – level this close collaboration with Eurodoc acted as a confidence builder on tackling gender expertise with a focus on ECRIs both for Eurodoc representatives as well as for the CoP facilitators. The CoP's enriched practice is also observed in the common vocabulary shared surrounding ECRIs. 'PhD researchers' came to replace the term 'students' as the perception of PhD researchers as part of the student body was progressively disrupted among the CoPs members.

Achievements and directions for future work

The work carried out within STRATEGIES in close collaboration with Eurodoc and many CoP members for whom the question of the challenges faced by ECRI felt meaningful, culminated with the preparation and submission of a successful proposal for a new COST[1] – (European Cooperation in Science and Technology) action set to be launched in October 2021. The COST Action is entitled 'VOICES: Making Early Career Investigators' Voices Heard for Gender Equality.'

We hope that through this COST Action which will last for four years beginning from October 2021, more instances (both individuals and institutions) will gain a better understanding of the fact that, although ECRIs constitute an important and fast-growing workforce, their working conditions remain precarious and their careers uncertain. These inequalities can be reinforced by disparities within academia linked to other social determinants, such as origin, socioeconomic status, sexuality, or ability which have been magnified by the recent COVID-19 crisis. With the launch of the action, six workgroups will be formed. As shown in Table 9.1, the Workgroups largely reflect the 5 thematic areas that are recommended for consideration in GEP according to Horizon Europe guidelines (European Commission, 2021).

The COST action will bring new life in the aftermath of STRATEGIES for Sustainable Gender Equality as a Community of Practice, albeit with a more specific topic, of the challenges faced by ECRIs. At the time of writing,

Table 9.1 Mapping between COST action CA 20137 workgroups and the content related building blocks of the Horizon Europe GEP eligibility criterion.

Thematic areas and GEP (as provided and requested in Horizon Europe)	*'VOICES, Making Early Career Investigators' Voices Heard for Gender Equality' workgroups*
Work-life balance and organisational culture.	Workgroup 1. Employment, Career Development and Mobilities
Gender equality in recruitment and career progression.	
Gender balance in leadership and decision-making.	Workgroup 2. Leadership and Decision Making
Integration of the gender dimension into research and teaching content.	Workgroup 3. Gender as a Research Dimension
Measures against gender-based violence, including sexual harassment.	Workgroup 4. Sexual Harassment, Gender Based Violence and Institutional Culture
	Workgroup 5. Intersectionality
	Workgroup 6. Monitoring and Evaluation

22 countries have joined the network. As the set-up and launch is still in progress, we hope that more countries will follow. Among them so far, we have also welcomed eight colleagues and institutions from the ACT project and STRATEGIES. In the meantime, a dissemination event is planned for October 2021 in the form of an open panel for the annual conference of the Society for Social Studies of Science on the same topic. The panel will welcome 11 contributions on three subtopics focusing on ECRIs: Institutional policies and initiatives, Intersectional discriminations in STEM and Precarity, Mobility, Work-Life Imbalances in Research Career Development.

Final reflections

The motivations and inner workings of STRATEGIES as a CoP reinforced our conviction that sustainability is a key concept for promoting gender equality and supporting institutional and organisational change at a national, European and international level. This chapter provided an overview of the theoretical and empirical underpinnings of the CoP.

In retrospect, we – as ACT CoP facilitators believe to have provided sufficient experiential evidence to demonstrate and argue that the strength of our CoP resided in the diverse knowledge and know-how acquired by the members of our CoP, the diverse experiences and needs manifested in different institutional and organisational contexts, the inclusive, bottom-up approach seeking to engage all perspectives from all CoP members and the common will to join forces and collaborate so as to create and make widely available the sharing of various resources, best practices and guidelines.

Despite the fact that our work has merely treated the tip of the iceberg 'sustainability in gender equality', we are very pleased that we were able to find a way to sustain the work that was initiated in our CoP. We believe that treating gender equality in a sustainable way will inevitably involve ruptures as much as (progressively more and more) continuities.

Note

1. COST is a funding organisation for research and innovation networks, funding actions which help connect research initiatives across Europe and beyond by enabling researchers and innovators to grow their ideas in any science and technology field by sharing them with their peers.

References

Aldercotte, Amanda, Guyan, Kevin, Lawson, Jamie, Neave, Stephanie, & Altorjai, Sylvia. 2017. ASSET 2016: Experiences of gender equality in STEMM academia and their intersections with ethnicity, sexual orientation, disability and age. Equality Challenge Unit. https://s3.eu-west-2.amazonaws.com/assets.creode.advancehe-document-manager/documents/ecu/ECUs-ASSET-report-November-2017_1579103163.pdf

Argyris, Chris, & Schön, Donald. 1978. *Organizational learning.* Addison-Wesley Pub. Co.

Blewitt, John. 2018. *Understanding sustainable development* (third edition). Routledge, Taylor & Francis Group.

Bozzon, Rossella, Murgia, Annalisa, & Poggio, Barbara. 2016. *Supporting Early Career Researchers through Gender Action Plans: A Design and Methodological Toolkit* (Vol. *9*). University of Trento.

Damala, Areti, van der Vaart, Merel, Van Dijk, Dick and der Sterke, Pam. 2021. Evaluation and impact assessment of emerging technologies: Institutional and organisational learning. In: Stylianou-Lambert T. and Shehade M., *Emerging Technologies and the Digital Transformation of Heritage Sites*, Springer, p. 55.

Esquivel, Valeria, & Sweetman, Caroline. 2016. Gender and the sustainable development goals. *Gender & Development, 24*(1), 1–8. https://doi.org/10.1080/13552074.2016.1153318

Eurodoc. 2020. *The aftermath of the pandemic for early career researchers in Europe.* (2020, July 6). Eurodoc. http://eurodoc.net/news/2020/the-aftermath-of-the-pandemic-for-early-career-researchers-in-europe

European Commission. 2012. *Communication from the commission to the European Parliament, the council, the European economic and social Committee and the committee of the regions. A Reinforced European Research Area Partnership for Excellence and Growth.* Retrieved from http://ec.europa.eu/research/science-society/document_library/pdf_06/era-communication-partnership-excellence-growth_en.pdf

European Commission. 2020. *Gender equality: Achievements in horizon 2020 and recommendations on the way forward.* Brussels, European Commission.

European Commission. 2021. *Horizon Europe guidance on gender equality plans.* Publications Office of the European Union. https://data.europa.eu/doi/10.2777/876509

Forest, Maxime & Lombardo, Emanuela. 2012. The Europeanization of gender equality policies: A discursive-sociological approach. In E. Lombardo & M. Forest (Eds.), *The Europeanization of Gender Equality Policies: A Discursive-Sociological Approach* (pp. 1–27). Palgrave Macmillan UK. https://doi.org/10.1057/9780230355378_1

Mihajlović Trbovc, Jovana, Reiland, Sonja, Reidl, Sybille and Damala, Areti. 2021. How to support Communities of Practice for driving institutional change towards gender equality. Policy brief, available at https://www.act-ongender.eu/sites/default/files/actongender_policy_brief_how_to_support_communities_of_practice_for_driving_institutional_change_towards.pdf#overlay-context=act-highlights

Morfaw, John. 2014. Fundamentals of project sustainability. *Paper presented at PMI® Global Congress 2014 – North America.* Phoenix, AZ. Newtown Square, PA: Project Management Institute.

Murgia, Annalisa & Poggio, Barbara. (Eds.). 2019. *Gender and precarious research careers: A comparative analysis.* Routledge.

Palmén, Rachel., Arroyo, Lidia, Muller, Jörg, Reidl, Sybille, Caprile, Maria, & Unger, Maximilian. 2020. Integrating the gender dimension in teaching, research content & knowledge and technology transfer: Validating the EFFORTI evaluation framework through three case studies in Europe. *Evaluation and Program Planning, 79*, 101751. https://doi.org/10.1016/j.evalprogplan.2019.101751

Sanders, Liz & Simons, George. 2009. A social vision for value co-creation in design. *Technology Innovation Management Review*. https://timreview.ca/article/310

Sautier, Marie. 2021. Move or perish? Sticky mobilities in the Swiss academic context. *Higher Education*, *81*(5). https://doi.org/10.1007/s10734-021-00722-7

Silvius, A.J. Gilbert & Schipper, Ron P.J. 2014. Sustainability in project management: A literature review and impact analysis. *Social Business*, *4*(1), 63–96. https://doi.org/10.1362/204440814X13948909253866

Tang, Catherine. 2002. Reflective diaries as a means of facilitating and assessing reflection. In *Quality conversations: Proceedings of the 29th HERDSA Annual Conference Perth* (pp. 7–10).

Thomson, Aleksandra, Rachel Palmén, Sybille Reidl, Sarah Barnard, Sarah Beranek, Andrew R. J. Dainty, and Tarek M. Hassan. 2021. Fostering collaborative approaches to gender equality interventions in higher education and research: the case of transnational and multi-institutional communities of practice. *Journal of Gender Studies*, *31*(1), 1–19.

Wallin, Patric, and Tom Adawi. 2018. The reflective diary as a method for the formative assessment of self-regulated learning. *European Journal of Engineering Education*, *43*(4), 507–521.

Wenger, Etienne. 2000. Communities of Practice and social learning systems. *Organization*, *7*(2), 225–246. https://doi.org/10.1177/135050840072002

Wenger, Etienne. 2011. *Communities of practice: A brief introduction*. https://scholarsbank.uoregon.edu/xmlui/handle/1794/11736

10 Benefits and limits of a CoP approach to promote gender equality in R&I

Sybille Reidl, Sarah Beranek and Florian Holzinger

Introduction

The European research and innovation (R&I) system is despite efforts and progress made still highly segregated by sex and gender, as evidenced by the latest edition of the She Figures and as already described in detail in the introductory chapter of this book. Gender equality is far from being achieved and much remains to be done to fulfil the three objectives of the ERA framework, namely (1) gender equality in scientific careers, (2) gender balance in decision-making processes, and (3) mainstreaming the gender dimension in the content of research and innovation (European Commission, 2019, p. 9).

To strengthen the engagement of Member States (MS) in promoting and supporting the establishment of the ERA, MS were asked to develop National Action Plans (NAP) that contain activities and efforts to progress towards reaching the objectives within the six priorities. For the gender equality priority, a recent monitoring report has detected that NAP activities focused on promoting gender equality led in some countries either to the first gender equality strategy for R&I (e.g., Cyprus, Luxembourg) or to the consolidation or further development of already existing policies that are at different stages of implementation (e.g., the Netherlands, Spain). Thus, the ERA process represents a tremendous potential, but as there are no incentives for relatively inactive countries to intensify their actions, "the gap between experienced and inactive countries with regard to gender equality in R&I is widening" (Wroblewski, 2021 p. 40). The ERA progress report draws a similar conclusion by stating "progress is slow and uneven across the ERA"[1] (European Commission, 2019 p. 10). Wroblewski (2021) also sees the risk that not all three gender dimensions are addressed equally by MS in their NAPs and consequently also on the European level.

In this chapter, however, the focus is not on the sociological mechanisms that explain why these inequalities exist and continue to be persistent, but on how to address them. Specifically, it discusses how Communities of

DOI: 10.4324/9781003225546-10

Practice (CoPs) can help to promote gender equality in the research system. In doing so, the chapter aims to answer the following questions:

- What added value do CoPs on gender equality have for their members?
- Is the CoP approach sufficient to foster gender equality in Research Performing Organisations (RPOs) and Research Funding Organisations (RFOs) regarding the ERA gender equality targets or are additional activities necessary?
- What are the limitations of this approach in terms of structural change for gender equality in R&I?

The chapter clearly shows the ways in which CoPs can support change efforts, what their limitations are, and what needs to be considered by policy makers in scaling up this approach at national and international levels. As CoPs have rarely been used as an instrument to promote gender equality so far, this is the first analysis of the extent to which CoPs are a helpful and instrumental tool for this purpose. This chapter provides, for the first time an understanding of the added values and benefits created through participation in inter-organisational CoPs and how this can be translated into effective changes in organisational practices. In addition, this chapter contributes to the discussion on how to promote gender equality in research and innovation and stimulate organisational participation in these change efforts.

To answer these questions, this chapter builds on the experiences and evidence from the Horizon 2020 project ACT. Its main aim was to promote gender equality within R&I and higher education (HE) organisations through inter-institutional exchange of gender practitioners from these organisations via the establishment of CoPs dealing with gender equality from different perspectives. In the project and in this chapter, CoPs are broadly defined as "groups of people who share a concern, a set of problems, or a passion about a topic, and who deepen their knowledge and expertise in this area by interacting on an ongoing basis" (Wenger, McDermott, & Snyder 2002, p. 4). In contrast to a network, in a CoP, there is a commitment of the individual organisations/actors to this relationship (Blackmore 2010), a shared identity and a collaboration on a joint project with a concrete result, as well as clear rules for membership, while networks tend to have more but looser connections (Holmes & Meyerhoff 1999). Thus while all CoPs qualify as networks, it does not work the other way around (Palmén et al. 2018).[2]

The endeavours of the CoPs took place within the three-dimensional gender equality concept of the ERA presented above and aim at a broad structural change approach. One approach to organisational change that is supported by the European Union and introduced as an eligibility criterion for Horizon Europe Funding are Gender Equality Plans[3] (GEPs). However, as the effective and efficient design, sustainable implementation and monitoring of GEPs is a highly complex matter, it presents many challenges, potential pitfalls and risks (e.g., resistance from within the organisation, lack of knowledge, resources, and support beyond the main gender equality agents) (EIGE 2016).

As gender inequalities are complex it can be difficult to achieve a broad understanding within an organisation, and so can be a challenge to bring about real organisational change within the framework of a GEP. The time-frame of a GEP adds here another layer of complexity (Kalpazidou Schmidt & Cacace 2017; Palmén & Kalpazidou Schmidt 2019). Thus, it might be ben-eficial to have the support of a CoP in organisational change towards gender equality. CoPs could be helpful in this process in many ways: Similar to GEPs, CoPs are built in an iterative process of mutual learning and capacity building, following the steps of audit, design, implementation and monitor-ing which might create synergies and the capacity to work collaboratively on the topic. In addition, CoPs have an identity-building effect among their members in terms of the issues they focus on, which could support the con-tinuity and sustainability of their gender equality efforts. Furthermore, in a CoP, a shared inter-organisational practice of exchanging resources, experiences, learnings, good-practices, methods and ideas takes place that produces know-how, strategies or even innovation potentially useful for putting GEPs into practice (Palmén, et al. 2018). The GenderTime project has already been able to show that CoPs are a valuable tool for support-ing structural change, as gender officers in particular, who were isolated in their institution, were able to benefit from the inter-organisational exchange and cooperation, e.g., by improving their competences (Barnard, Hassan, Dainty, Polo, & Arrizabalaga 2016). However, CoPs are not limited to the topic of GEPs, since not all institutions might be in the right time and place (i.e., context) to tackle such a complex undertaking. In any case, CoPs are also suitable to support any form of gender-sensitive activities, as the ACT results will show.

The ACT project featured seven CoPs[4] that all dealt with the topic of gen-der equality in the research system. While all of the CoPs were based upon an inter-organisational cooperation, thus uniform in this aspect, they were, however, quite heterogeneous in their thematic foci. The seven ACT CoPs forming part of the evaluation were:

- LifeSciCoP – Gender Equality in Life Sciences,
- GEinCEE – Gender Equality in Central and Eastern Europe,
- GenBUDGET – Gender Budgeting in Research Organisations,
- FORGEN – Funding Organisations for Gender,
- GENERA – Gender Equality in Physics and beyond,
- STRATEGIES – Strategies for Sustainable Gender Equality,
- Alt+G – Alternative Infrastructure for Gender Equality

The CoP facilitators, who were also part of the ACT consortium, started the CoPs, but the topics, approaches and goals emerged through the discussions of the members. In spring 2020, the project had 132 CoP members from 26 countries[5]; about half of them were from universities, about a third from Research Organisations, 14% from Funding Organisations and 4% from Associations/Federations of universities or research institutions.

During the project lifetime, it was foreseen that the people who participate in the ACT CoPs commit to the CoP, as well as engage in learning and exchange about organisational and inter-organisational gender equality work. In order to achieve change within R&I organisations, however, this learning has to be taken up and fed into the organisations of the members. Thus, the learning process starts within the collaboration of the CoP and ideally unfolds in the organisation. However, how does learning work in a CoP setting? While we do not address the workings of organisational learning (Pawlowsky & Geppert 2005) here, it is evident that the strength of the CoP approach lies in its knowledge sharing and collaborative learning intentions. In the CoP approach, the underlying understanding is that learning is not primarily based on the cognitive processing of the individual members in the form of receiving factual knowledge or information, but on social processes (Lave & Wenger 1991). This means that members do not simply pass on their knowledge to the other members, as would be done with a manual. Instead, members learn through interaction and participation and construct their knowledge in a way of acting that is situated in a particular time and (culturally located) system (Blackler 1995). Through their group-character, CoPs can construct inter-subjective meaning and with it a shared reality: "Words, labels, metaphors, and platitudes produce the reality that people experience as 'out there'" (Gherardi & Nicolini 2001). Consequently, CoPs can be considered highly flexible, effective and innovative in responding to the needs of members as they constantly adapt to changing membership and changing circumstances (Brown & Duguid 1991). Problems and questions, which, for example, arise in the context of the organisational gender equality work, can be discussed directly when they emerge in practice and solved in a collaborative way and "out of this friction of competing ideas can come the sort of improvisational sparks necessary for igniting organisational innovation" (Brown & Duguid 1991, p. 56). Accordingly, this learning process, which extends from the individual member's participation in the CoP to the respective R&I organisation is, thus, not understood in a linear way, but as a cycle leading back to the CoP.

How can this learning process lead to organisational change? Similar to the process of learning in a CoP, the approach of achieving organisational change, i.e., regarding gender equality in the respective organisation, is considered a form of doing – a practice-based approach. As the CoP provides a different setting in terms of hierarchies and routines than the respective organisation, the members have the possibility to experiment with alternative ways of doing, thus exploring new practices. Change is, thereby not primarily about new information, but on remodelling established practices (Bruni, Gherardi, & Poggio, 2004; Gherardi, Cozza, & Poggio 2018), which can be easier in new communities than in already existing ones (Roberts, 2006). Remodelling can be initiated by "practical reflexivity" (Cunliffe & Easterby-Smith, 2004), hence, "the collective process of reflecting on taken for granted forms of practising and bringing them at a discursive level open [for] new possibilities of actions" (Gherardi 2015). However, then, these practices have to be integrated into the respective organisational setting as they cannot be directly transferred from organisation to organisation: "For practitioners to be able to recognise 'a shared way of doing things', and therefore for practices to work as practical and temporary agreements on how action should be carried on, it is necessary for a practice to be institutionalized even when its institutionalization is contested or challenged." (Gherardi 2009, p. 356).

Thus, it is evident that the strength of the CoP approach lies in its knowledge sharing and collaborative learning intentions. Due to its practical relevance, this type of learning seems to be well suited for the gender field (Ostermann 2003). There has been quite some research on the success factors, barriers and dysfunctions of CoPs (Gelin & Milusheva 2011; Hammer, Beck, & Glückler 2012; Mládková 2015; Pyrko, Dörfler, & Eden 2017; Vincent, Steynor, Waagsaether, & Cull 2018), as well as on CoP benefits (Fontaine & Millen 2004; Millen, Fontaine, & Müller 2002; Zboralski, Salomo, & Gemuenden 2007), or both (Gannon-Leary & Fontainha 2007). This chapter joins the still rather limited research on CoPs in the field of gender equality in research (Barnard et al. 2016), but more concretely on CoP benefits and limitations in this specific area. Hence, this chapter will present the benefits and learnings that the members could create while participating in the ACT CoPs. When looking at benefits, the answers will be structured in alignment with the concept of Wenger (2011) and its cycles of value creation: (1) immediate value, (2) potential value, (3) realised value, (4) applied value, (5) reframing value.

Materials and methods

The empirical evidence of this chapter is based on evaluation activities[6] within the ACT project. The evaluation aimed to (1) identify challenges and strategies for developing CoPs for institutional change, (2) assess the usefulness of the learning outcomes for the CoPs, as well as (3) the effect of the CoPs on the development of gender equality in their member organisations, and (4) develop recommendations on how CoPs need to be designed to be more effective for their members in promoting gender equality. Two evaluation instruments were used to gather the evidence needed to answer the evaluative questions regarding added value and benefits[7]:

Semi-structured interviews with CoP facilitators (n = 7) and CoP members (n = 21): The interviews (30–90 min) took place between May and July 2020 and were conducted via video call by the project partner JOANNEUM RESEARCH. The three interviewed members per CoP were selected following criteria that aimed at getting a diverse picture (e.g., in terms of their organisation size, region/country, or number of CoP meetings attended). Some of the interview partners were gender equality officers or working in the Human Resources departments, however, the majority were researchers and/or mostly women (86%). The member interview questions targeted the personal background of the interview partner, their participation in CoP activities, as well as cooperation and communication with other members and their perspectives on the benefits and impact of CoP involvement. Furthermore, they were asked about the perceived limitations of the CoP approach and further needs to achieve structural change in R&I and their organisation. The facilitator interview guide addressed similar areas, but from a different perspective. The interviews were recorded and transcripts were generated in a summative form.

Progress reports: The progress reports were written by the CoP facilitators and document the development of the CoPs based on the monitoring that the CoP facilitators continuously filled in. They contain detailed information on the CoPs' objectives, composition, activities and progress, as well as providing glimpses into their working space via meeting protocols, social media content such as blogposts or tweets. The interview transcripts and progress reports were analysed with codes that were developed deductively based on the evaluative questions via the software MAXQDA. Codes that emerged inductively during coding or analysis were also integrated. Thus, the process included several iterations of coding and analysis.

Results on CoP value creation

Based on the results of the analysis described above, this chapter presents answers to the following questions: How do members benefit from the CoP? Which learnings can be identified? What are the effects of CoPs at the organisational level regarding implementation of gender equality activities? Where are the limits of the CoP approach? And, what is required beyond the CoP activities to enable members to foster structural change regarding GE successfully?

To analyse the learning outcomes from CoPs and their effects on gender equality development in the members' organisations, we took the concept of Wenger et al. (2011) about promoting and assessing value creation in communities and networks. Wenger et al. distinguish five cycles of value creation: (1) immediate value, (2) potential value, (3) applied value, (4) realised value, (5) reframing value.

1 *Immediate value.* The first form is the most fundamental cycle. It considers activities and interactions in networks/communities and interactions as valuable in themselves (Wenger et al. 2011, p. 19). In ACT e.g., the CoP meetings gave members the opportunity to ask colleagues to provide tips to solve a difficult case, passing information along, get inspirations from others who talk about their experiences, etc. We can see this "immediate value" as direct, quickly collected small benefits,

which members of all ACT communities could gain when participating in CoP meetings but also in Mutual Learning Events that took place in all of the seven evaluated CoPs focussing on different topics (GEP implementation, resistance, Gender Equality and Audit Monitoring Survey (see Chapter 3 this volume), gender in COVID-19 funded research and effects of COVID-19 on researchers, evaluation and impact assessment, gender budgeting and gender equality in science and research).

2 *Potential value.* Not all the value produced by a community is realised immediately. Activities and interactions can produce "knowledge capital," the value of which lies in its potential for later realisation. For example, a successful strategy against resistance to gender equality in top management can be learned from a CoP member's experience. Even if you never face this resistance yourself, it is useful and reassuring to have this knowledge just in case (Wenger et al. 2011, p. 19). This knowledge capital can take different forms:

- Personal change through CoP membership
- Change in social relationships
- Access to resources
- Acquired position of the CoP

Through analysing the interviews with CoP members and facilitators, we can identify value creation in the ACT CoPs in all these four forms of potential value:

a *Personal change through CoP membership (human capital).* The CoP members mentioned different personal benefits of their participation in the CoPs: They exchange experiences, strategies with implementing GE measures, which is needed "to start a change process in one's own organisation" as a member (M1) of GEinCEE puts it. Members also report awareness raising because they learnt about representation of women in academia, different national and organisational situations and about gender equality policies at EU level, as CoP facilitators report (F6, F7). Especially members with little gender equality experience had "aha-effects": "I learnt that what I thought was the problem is just the tip of the iceberg!" (GENERA M3). To increase their knowledge, members also got inputs from external experts. Moreover, CoP members benefited from interdisciplinary cooperation in the CoP, which opened up new perspectives. They gained confidence in their ability to engage in practice and felt inspired for their GE work: "Working in the working groups (of the CoP) has really been an inspiration for me to do more!" (GENERA M3).

b *Access to resources (tangible capital).* To further develop the GE knowledge of their members, CoPs built sound knowledge bases in sharing research results, literature, links and interesting studies etc. about issues they are working on (e.g., sexual harassment, resistance but also methodological knowledge like gender budgeting and GE monitoring). They also provide their members with methodological tools (GEAM Tool, Co-Creation Toolkit, social science methods) that they can use in their efforts to promote structural change within their organisations: "the main thing we are working now is the GEAM survey" (GEinCEE Member 3). Two CoPs also offer financial support for participation in face-to-face meetings (as they were relevant before the COVID-19 pandemic).

c *Change in social relationships (social capital).* The CoP members in ACT also benefit socially from their participation. Through the CoP, they got access to new people and expand their social network, especially in disciplines and regions with which they do not otherwise come into contact easily. With a continuous participation in the CoP the members establish trust, perceive the CoP as a safe space and know who to turn to for help and feel less isolated: "A sense of community and that people want to help, they take time to help you, reply to you in case of need (...). So this is really, it makes you sleep better, I think, knowing that there are people who are not selfish." (Alt+G M3).

d *Acquired position of the CoP.* In addition to the sense of community and personal support, CoP members also gain a new voice to the outside world through the CoP. "When I give a statement in the media supported by the network, this gives it more weight" (Alt+G M3). The CoP increases the visibility and awareness of GE and its members. In addition, CoPs are needed to foster societal change towards gender equality: "The CoP helps with long term fights like gender sensitive language – you cannot do that alone, this needs many people from different institutions who push that" (Alt+G M2).

3 *Applied value.* When CoP members start to make use of what they gained in the CoP, we talk about applied value. For example, knowledge capital is a potential value that may or may not be realised. The use of knowledge requires adaptation and application to a specific situation (e.g., adapting the GEAM survey, apply new methods in internal projects). Adapting and applying knowledge capital in different contexts can lead to changes or innovations in actions, practice, tools, approaches, or organisational systems. Wenger calls this applied value – the third cycle of value creation (Wenger et al. 2011, p. 21) and identifies different forms that we also encountered in the evaluation of ACT CoPs:

- Use of experiences and knowledge
- Use of CoP products
- Attracting colleagues to join
- Implementation of gender equality activities

a *Use of experiences and knowledge.* CoP members distribute experiences and knowledge gained in the CoP to gender equality committees, diversity teams, etc. in their organisation: "I will share the idea about the equity committee with my colleagues" (FORGEN M3). Organisational data collected in a long-term monitoring of gender relevant data of CoP member organisations in GENERA is supportively used in advocating with the management. Members of GenBUDGET feed their learnings about gender budgeting directly in internal projects e.g., about improving salary conditions for employees.

b *Use of CoP products.* At the time of the interviews, most CoPs planned to develop concrete products such as an online gender equality map (GEinCEE) or a GEAM survey module for RFOs (FORGEN). GEinCEE and LifeSciCoP member organisations were already using the GEAM survey, which had been translated by the CoPs and supplemented with questions on COVID-19: "The GEAM survey is the most important initiative of the CoP because the results show us the changes in the university we can develop, it shows us the direction." (GEinCEE M2). For all other CoPs, the products were in the development phase at that time, so it was too early to talk about their use.

c *Attracting colleagues to join.* Some CoP members are researchers who are interested in gender equality in their organisation, and are very much alone in this. The CoP can support them to find allies, if their members succeed in getting colleagues from their environment interested in the topic. Some members of ACT CoPs have already started to involve colleagues and students in activities by inviting them to CoP workshops or meetings. In one organisation, this led to starting an initiative group to foster gender equality: "We are organising conferences, we are working with the GEAM tool, we planned and developed a GEP. Something started actually with the presence of the CoP to stimulate activity." (GEinCEE M3) Another member of GEinCEE succeeded in interesting representatives of the management to form an implementation team (GEinCEE M1). In GenBUDGET, a member formed a working group to conduct their targeted implementation project and also managed to involve HR and the data protection officer (M2).

d *Implementation of gender equality activities.* These examples show that already after about a year of existence, CoPs stimulate initial activities to promote gender equality in some member organisations. At this early stage, we can only report initial anecdotal evidence regarding implementation of gender equality activities, such as the introduction of gender-inclusive language or the implementation of a gender monitoring to enable the pinpointing of problems to the management: "In some organisations they could convince the local faculty to do gender inclusive language" (Alt+G M2).

4 *Realised value.* It is still too early to be able to show realised value – the fourth circle of value creation according to Wenger (2011). At this stage, effects of applied knowledge capital to achieve equality can be identified. So far, we can only detect first signs: Sporadically, CoP members were able to get their management to discuss gender equality issues in the organisation. In general, visibility of gender equality issues increased in member organisations due to CoP participation, as several CoP facilitators report. What effects this will have in the future remains to be seen.

5 *Reframing value.* In the final cycle of value creation according to Wenger et al. (2011), social learning causes a rethinking of learning imperatives and the criteria by which success is defined. This form of value creation has not yet been achieved by ACT CoPs.

Limitations of the CoP approach

After shedding light on the question of value creation through the CoP, we will now turn to the question if this CoP approach is sufficient to foster gender equality in RPOs and RFOs and contribute to reaching the ERA targets.

The ACT CoPs have succeeded in developing common goals and an incipient group identity in the course of their existence. Members derive their personal benefits from membership in the CoP and the first signs of organisational activities are emerging (individual members conduct status quo analyses, negotiate with management on the implementation of gender equality measures, intra-organisational gender equality teams are established, etc.). To achieve the ERA goals, the European Commission sees the development and implementation of GEPs in RPOs and RFOs as a proven means. In this respect, the original goal of CoPs envisioned in ACT was to support members in this process. In practical work with CoP members, however, it soon became apparent that this goal was too ambitious for many members.

For several members in different CoPs participating in the CoP is not part of their job description and therefore voluntary and unpaid work on top of their daily work. They are researchers, interested in advancing gender equality in their organisation but have no mandate to do so and are often alone in their concern to advance gender equality in their organisation. Their time to invest in gender equality activities is restricted and much more time resources would be needed to develop GEPs, conduct a GEAM survey, monitor, and form part of gender equality committees, etc. As the resources of some CoP members are extremely limited, it is very difficult to strive for long-term and comprehensive equality goals such as cultural change or the design and implementation of a GEP. As a first step, their organisation would first need to become aware of the need for gender equality goals and foresee resources for them. The CoP approach can support gender activists, but it needs to be complemented by institutional buy in: "A CoP hopefully is a support structure for being activist, having enough knowledge, being pushy but receptive to others' world view at the same time to do that human interaction that might slowly change culture. But the CoP alone is not enough" (GenBUDGET M3). This kind of organisational change process might also need specific consulting within the organisation:

> The CoP is good for exchange of knowledge, benchmarking against each other and getting inspiration, but then we would need concrete support in our organisation for a change process – that cannot be achieved only with exchange in the CoP.
>
> *(LifeSciCoP M3)*

A CoP can provide inspiration, (tacit) knowledge and methods, support in developing strategies how to start activities, help to overcome resistance but: "a CoP can never be everything, it can ... complement, [be] something helpful" (GenBUDGET M1).

The European Commission (EC 2011) has highlighted key elements for effective institutional change:

- knowing the institution: collecting base-line data at the institutional level in order to inform an institutionally tailored, evidence-based strategy of change;
- securing top-level support: without commitment by leadership, institutional change lacks strategic importance, and implementation may just not happen, be circumvented, or resisted;
- effective management practices: raising awareness and building gender competence of key decision-makers, human resource managers and other relevant stakeholders

Linked to the top-level support, EIGE (2016) highlights the availability of adequate resources as a prerequisite for institutional change. However, resources are also necessary to collect organisational data and raise awareness of management. The gender equality agenda and the implementation of GEPs rely not only on the individual agency and activism of a few passionate individuals in institutions often with no collective force or voice to systematically challenge the status quo, who are supported in CoPs. Institutional change also relies on the structure of stable governance frameworks, legislation, resources, and external incentives (Palmén and Kalpazidou Schmidt 2019). These forms of support are also requested by various CoP members to speed up organisational change. Especially CoP members from Eastern Europe report particularly difficult framework conditions in this respect. They face massive resistance against GE and an overwhelming amount of hate:

> We can support each other but this ends where the social reality starts. We don't know how to fight this pure hate. I think the limits of the CoP are in the reality and political atmosphere, which is created by the government.
>
> *(GEinCEE M2)*

Moreover, CoP members mention the need of support from strategic actors like policy makers or funders on the national level to get more resources for GE. Therefore, some of them try to acquire resources through structural change projects funded by the European Commission (GEinCEE M3).

Discussion

Overall, we see in our analysis that CoPs generate different forms of values – depending on the duration of their existence but also on the duration of membership, starting at the personal level as well as at the levels of knowledge capital and social capital. This corresponds with findings in the literature: Pyrko et al. (2017) show in their analysis the importance of immediate value to justify the members' time investment when they start joining a CoP. They describe this immediate value as: engaging discussions, new working relationships, ability to share their views, solutions to their problems, opportunities to see what others are doing, and some tools, documents or techniques that they could use in their work. We can see many parallels to our findings in the ACT CoPs: e.g., in meetings, members heard about GEP implementation, learnt how others deal with resistance or were introduced to useful tools that can be used to assess the status quo of gender equality in an organisation.

These short-term benefits are important for people to remain and decide to consistently engage with the CoP, because participation is voluntary and often also without remuneration and recognition. Nevertheless, the full value of a CoP is often not apparent at the beginning and the source of value often changes over the lifetime of the CoP. Early value often comes from focusing on the current problems and needs of community members. Whereas, later on developing a systematic body of knowledge that can be easily accessed becomes more important (Wenger et al. 2002). In individual ACT CoPs, this was, for example, the collection of intervention options for gender equality in the research funding process or the further development of the GEAM tool.

However, it is not only about gender equality knowledge. EIGE (2016) shows that in addition to a lack of gender equality knowledge, gender practitioners struggle primarily with resistance, insufficient management support, lack of resources and lack of authority and decision-making power to develop and implement GEPs (see also Bleijenbergh 2018; Verge, Ferrer-Fons, González 2018; Palmén, Kalpazidou Schmidt 2019). This is particularly true in many Central and Eastern European countries as discussed in Chapter 5 of this volume by Sekula et al. Sufficient resources, gender expertise, awareness and competence among the organisational stakeholders to design and implement gender equality interventions and decisive power of gender equality bodies are important facilitators of effective gender equality interventions (EIGE 2016; Lansu, Bleijenbergh, Benschop 2019; Lipinsky 2014; Mergaert, Lombardo 2014; Palmén and Kalpazidou Schmidt 2019).

CoPs can not only help to improve the gender knowledge of their members and increase their awareness of the need for more gender equality and their competences for implementing gender equality measures. They also help to jointly develop strategies how to convince the management of the need for gender equality and acquire more resources, etc. Learning is experienced here in a very practical way.

Gannon-Leary & Fontainha (2007) emphasise in this context the enhanced learning environment in a CoP. Learning in CoPs is collaborative. It is based on collaborative knowledge of the community, which is greater than any individual knowledge (Johnson 2001) and was also often mentioned by CoP members as an important advantage of the CoP. Learning in CoPs is based on the concept of sharing. Members can thus continuously deepen their knowledge and expertise. Gannon-Leary & Fontainha (2007) thus describe knowledge development in a CoP as continuous, cyclical, and fluid, with no clearly defined beginning or end. This "networked" or "situated" learning focuses not on abstract bodies of knowledge but on engagement with real-world problems. CoPs encompass this concept in that they establish a networked environment where the necessary interactions for improving learning can occur (Wenger et al. 2002). The interactions focus around knowledge sharing among members, who may range from experts through to novices, and can also take the form of a neo-apprenticeship style learning. Subsequently, this type of learning enables a practice-based use of the acquired knowledge and a sense of connection and group identity. In the ACT CoPs, members moreover benefited from gaining new perspectives on implementing gender equality measures from different stakeholders like gender equality practitioners, researchers from various disciplines, human resource managers, and top management. This resonates with Hearns and Whites (2009) findings that CoPs become a platform for bringing together different types of competences. Furthermore, the evaluation of ACT CoPs shows that CoPs have a social impact on their members, as they counteract social isolation that gender activists often experience when they are actively engaged in gender equality solely in their organisation (Barnard et al. 2016). ACT CoPs support their members by helping to combat this isolation through networking with gender activists from other organisations. However, to be successful in the long term within the organisation, gender activists also need an internal network of allies. CoPs can support the search for these allies by offering interesting workshops for those interested, discussing strategies how to find allies, etc.

Over time, CoPs can also establish a process of "thinking together," which Pyrko et al. (2017, p. 389) identify as a key part of meaningful CoPs "where people mutually guide each other through their understandings of the same problems in their area of mutual interest, and this way indirectly share tacit knowledge." In the ACT CoPs, we can identify these kinds of processes in various forms, for example when, the FORGEN CoP starts to work together on the grant evaluation process or GEinCEE collaborates in conducting status quo assessments in their member organisations, but also when starting to develop strategies e.g., for convincing the top management, as mentioned before. This "thinking together" also goes beyond joint activities such as those described here. For example, we see in the Alt+G CoP how a CoP can lobby for gender equality issues such as gender-sensitive language at the national level. Over time, CoPs become more visible to the outside world and can also contribute to the legitimacy of gender equality in their member organisations and can play a role of intermediate support structures that connect gender equality activities and strengthen conditions necessary for structural change (see also Chapter 5). CoPs can also force benchmarking between their members by comparing institutional data on gender equality. Members can use this as an argument for gender equality activities to the top management of their organisation. CoPs become a relevant stakeholder in their environment driving gender equality issues. This can subsequently help to ensure high level support in organisations – a key factor for successful GE implementation, according to Palmén and Kalpazidou Schmidt (2019).

With prolonging these activities, the CoPs might also deliver value to the organisation and to the teams on which community members serve. They might promote cooperation between organisations and knowledge transfer, thus contributing to innovation (Bertels, Kleinschmidt, Koen 2011) and stimulate open innovation, which facilitates enhanced knowledge acquisition and transfer (Pattinson, Preece, & Dawson 2016) and might also lead to structural change towards gender equality and foster gendered innovations. But at the time the evaluation was conducted, it was still too early to empirically prove the direct organisational value. Moreover, Millen et al. (2002) identify organisational benefits of CoPs that involve improved communication among community members, which contributed to successfully executed projects; time savings in information seeking that contributed to improved efficiency. For CoPs in the field of gender equality, one could also imagine that they could support transparent processes in the organisation (e.g., in the field of recruitment or funding) or lead to more excellent research if they succeed in contributing to organisational change towards gender equality in the long run.

Measuring and demonstrating the value of CoP is a difficult endeavour (Millen et al. 2002). Wenger et al. (2002) show that one must consciously go in search of the added value of the CoP, that one must first train one's perception of the benefits generated here in order to substantiate the diffuse feeling that membership in the CoP is useful, with hard facts: "Many of the most valuable community activities are the small, everyday interactions—informal discussions to solve a problem, or one-on-one exchanges of information about a tool, supplier, approach, or database. The real value of these exchanges may not be evident immediately" (Wenger et al. 2002, p. 60).

Conclusion

Our findings show that the CoP approach can contribute substantially to support passionate individual change agents and activists who are needed to start and drive institutional change towards GE. As Thomson et al. (2021) put it, CoPs "supported them through a much-needed collective identity and a sense of belonging, where knowledge and best practices could be exchanged. CoPs offered a retreat to seek mutual appreciation of similar problems related to limited resources, strategies against resistance, and a 'safe space'." To be able to fulfil this very important role, CoPs also need resources to sustain themselves. To establish an effective network and make substantial progress in the member organisations and activities started, sustainable resources are needed that go beyond temporary project funding.

Overall, the evaluation of ACT CoPs shows that CoPs are a piece of the puzzle when it comes to organisational change towards gender equality. They are an important support for change agents in organisations and contribute to the visibility of GE also at national level (see also Chapter 5). However, they have no possibilities to counter the lack of resources for implementing gender equality except lobbying activities. In the design and implementation of GEPs in organisations, structural change projects funded by the European Commission have so far played a central role that CoPs cannot compensate for. We were also able to show that CoPs can lend weight to the issue of gender equality in organisations. However, top management can be even more convinced of the need for action if national and international legislation and regulations exert pressure towards the implementation of gender equality, as is currently the case with GEPs as eligibility criterion for Horizon Europe.

Moreover, a concept for scaling up is needed to enable as many organisations as possible to become members of a CoP. CoPs are limited in size. In order to remain operational, they cannot be enlarged to any size. In GENERA – the largest CoP with 34 institutional members – CoP facilitators are therefore now carefully considering who else to accept as members. To address the various regional and thematic challenges – starting with the three ERA priorities – gender equality activists throughout Europe are facing, a broader offer of CoPs is needed than the ACT project was able to provide.

Notes

1. While both reports agree on this point, they come to different results regarding the development individual countries. According to Wroblewski 2021, the ERA progress report would benefit from using a wider and more differentiated set of indicators.
2. For a more detailed description of the (ACT) CoP approach, as well as the tension field between theoretical definition and practice of CoPs, see the Introduction of this book.
3. The European Commission (2012) defines a GEP as a plan "aiming at: (1) Conducting impact assessment / audits of procedures and practices to identify

gender bias; (2) Identifying and implementing innovative strategies to correct any bias; (3) Setting targets and monitoring progress via indicators."

4. As this chapter is based on the evaluation results, no statements are made about the 8th CoP LAC (Gender equality in Latin America), as they were added later to the project and thus were not part of the evaluation.
5. When including the LAC CoP members, the ACT project support in total 144 organisations across eight Communities of Practice.
6. It is confirmed by the Head of the Ethics Board at JOANNEUM RESEARCH, the partner responsible for conducting the research, that all the research processes conducted for the purposes of this article meet the JOANNEUM RESEARCH and the GDPR standards. All participants signed an informed consent about the purpose of the study.
7. The evaluation also included a quantitative survey called the Wilder Collaboration Inventory Questionnaire, which was adapted to the specifics of the ACT project (Mattessich, Murray-Close, & Monsey 2001). However, as the questionnaire did not address the added value of CoP participation, it will not be presented in more details in this chapter.

References

Barnard, Sarah, Hassan, Tarek, Dainty, Andrew, Polo, Lucia, Arrizabalaga, Ezekiela (2016). Using communities of practice to support the implementation of gender equality plans: lessons from a cross-national action research project. Loughborough University. Conference contribution. https://hdl.handle.net/2134/23681

Bertels, Heidi M. J., Kleinschmidt Elko J. & Koen Peter A. (2011). Communities of practice versus organizational climate: Which one matters more to dispersed collaboration in the front end of innovation? *Journal of Product Innovation Management 28(5)*, S. 757–772. doi: 10.1111/j.1540-5885.2011.00836.x

Blackler, Frank (1995). Knowledge, knowledge work and organizations: An overview and interpretation, *Organization Studies 16(6)*, 1021–1046. doi: 10.1177%2F017084069501600605

Blackmore, Chris (2010). *Social learning systems and communities of practice.* London, New York: Springer. doi: 10.1007/978-1-84996-133-2

Bleijenbergh, Inge (2018). Transformational change towards gender equality: An autobiographical reflection on resistance during participatory action research. *Organization 25(1)*, 131–138. doi: 10.1177/1350508417726547

Brown, John S. & Duguid, Paul (1991). Organizational learning and Communities of Practice: Toward a unified view of working, learning, and innovation. *Organization Science 2(1)*, 40–57. doi: 10.1287/orsc.2.1.40

Bruni, Attila, Gherardi, Silvia, & Poggio, Barbara (2004). Doing gender, doing entrepreneurship: An ethnographic account of intertwined practices. *Gender, Work and Organization, 11*(4), 406–429. https://doi.org/10.1111/j.1468-0432.2004.00240.x

Cunliffe, Anne & Easterby-Smith, Mark (2004). From reflection to practical reflexivity: experiential learning as lived experience. *Organizing Reflection.* Aldershot: Ashgate Publishing, pp. 30–46.

EIGE. (2016). *Roadmap to Gender Equality Plans in research and higher education institutions.* https://eige.europa.eu/sites/default/files/gear_roadmap_02_success factors_obstacles.pdf

European Commission, Directorate-General for Research and Innovation (2019). ERA progress report 2018: the European Research Area : advancing together the Europe of research and innovation, Publications Office, 2019, https://data.europa. eu/doi/10.2777/118067

Fontaine, Michael, & Millen, David R. (2004). Understanding the benefits and impact of Communities of Practice. In Paul Hildreth, & Chris Kimble (Eds.), *Knowledge Networks: Innovation Through Communities of Practice* (S. 1–13). doi: 10.4018/978-1-59140-200-8.ch001

Gannon-Leary, Pat & Fontainha, Elsa (2007). Communities of Practice and Virtual Learning Communities: Benefits, Barriers and Success Factors. *elarning Papers No. 5*, September 2007, Available at SSRN: https://ssrn.com/abstract=1018066

Gelin, Philippe, & Milusheva, Maia (2011). The secrets of successful communities of practice: Real benefits from collaboration within social networks at Schneider Electric. *Global Business and Organizational Excellence 30(5)*, S. 6–18. doi: 10. 1002/joe.20391

Gherardi, Silvia (2009). Knowing and learning in practice-based studies: An introduction. *The Learning Organization*, special issue *(16)*, 5. doi: 10.1108/ 09696470910974144

Gherardi, Silvia (2012). Why do practices change and why do they persist? models of explanations. In Paul Hager, Alison Lee, Ann Reich (Eds.), *Practice, Learning and Change: Practice Theory Perspectives on Professional Learning.* New York: Springer International. doi: 10.1007/978-94-007-4774-6

Gherardi, Silvia (2015). How the turn to practice may contribute to working life studies. *Nordic Journal of Working Life Studies 5*, 13–25. doi:10.19154/njwls. v5i3a.4831

Gherardi, Silvia, Cozza, Michela & Poggio, Barbara (2018). Organizational members as storywriters: on organizing practices of reflexivity. *The Learning Organization 25*, 51–62. doi: 10.1108/TLO-08-2017-0080

Gherardi, Silvia, & Nicolini, Davide (2001). The sociological foundations of organizational learning. In Meinolf Dierkes, Ariane Berthoin Anthal, John Child, & Ikujiro Nonaka, *Handbook of Organizational Learning and Knowledge (S. 35–60).* Oxford: Oxford University Press.

Hammer, Ingmar, Beck, Silke, & Glückler, Johannes (2012). Lernen im lokalen Unternehmensnetz-werk. Imitation zwischen Konvention und Tabu. In Johannes Glückler, Waltraud Dehning, Monique Hearn, Simon, White, Nancy. *Communities of Practice. Linking knowledge, science and policy.* ODI Background Note. ODI: London, p. 4, https://cdn.odi.org/media/documents/1732.pdf

Janneck, Monique & Thomas Armbrüster (Eds.), *Unternehmensnetzwerke. Architekturen, Strukturen und Strategien* (S. 163–185). Berlin, Heidelberg: Springer Gabler. doi: 10.1007/978-3-642-29531-7_9

Hearn, Simon, & White, Nancy (2009*). Communities of practice: Linking knowledge, policy and practice.* Background Note. Overseas Development Institute.

Holmes, Janet, & Meyerhoff, Miriam (1999). The Community of Practice. Theories and methodologies in language and gender research. *Language in Society 41(2)*, S. 173–183. doi: 10.1017/S004740459900202X

Johnson, Christopher M. (2001). A survey of current research on online communities of practice. *The Internet and Higher Education 4(1)*, 45–60. doi: 10.1016/ S1096-7516(01)00047-1

Kalpazidou Schmidt, Evanthia, & Cacace, Marina (2017). Addressing gender inequality in Science: The multifaced challenge of assessing impact. *Research Evaluation 26(2)*, 589–617. doi: 10.1093/reseval/rvx003

Lansu Monic, Bleijenbergh, Inge, Benschop Yvonne (2019). Seeing the system: Systemic gender knowledge to support transformational change towards gender equality in science. *Gender Work & Organisation 26(11)*, 1589–1605. doi: 10.1111/gwao.12384

Lave, Jean, Wenger, Etienne (1991). *Situated Learning: Legitimate Peripheral Participation*. Cambridge University Press.

Lipinsky Anke, (2014) Gender equality policies in public research. *Based on a Survey among Members of the Helsinki Group on Gender in Research and Innovation 2013*, Luxembourg: European Commission, https://op.europa.eu/en/publication-detail/-/publication/39136151-cb1f-417c-89fb-a9a5f3b95e87, 06.10.2021.

Mattessich, Paul, Murray-Close, Marta, & Monsey, Barbara (2001). *What Makes it Work*. St. Paul, MN: Amherst H. Wilder Foundation.

Mergaert, Lut, Emanuela Lombardo (2014). 'Resistance to implementing gender mainstreaming in EU research policy', In: Weiner, Elaine and Heather MacRae (Eds.) *The persistent invisibility of gender in EU policy' European Integration online Papers (EIoP)*, Special issue 1, Vol. *18*, Article 5.

Millen, David R., Fontaine, Michael A., & Müller, Michael J. (2002). Understanding the benefit and costs of communities of practice. *Communications of the ACM 45(4)*, S. 69–73. doi: 10.1145/505248.505276

Mládková, Ludmila (2015). Dysfunctional Communities of Practice – Thread for organization. *Procedia – Social and Behavioral Sciences 210*, S. 440–448. doi: 10.1016/j.sbspro.2015.11.392

Ostermann, Ana Christina (2003). Communities of practice at work: Gender, face-work and the power of habitus at an all-female police station and a feminist crisis intervention in Brazil. *Discourse & Society 14(4)*, S. 473–505. doi: 10.1177/0957926503014004004

Palmén, Rachel, Caprile, Maria, Panadès, Rosa, Riesco, Julia, Pollitzer, Elizabeth, & Vikenburg, Claartje (2018). ACT Conceptual Framework. https://www.gender portal.eu/resources/act-conceptual-framework. doi: 10.5281/zenodo.3235296

Palmén, Rachel, & Kalpazidou Schmidt, Evanthia (2019). Analysing facilitating and hindering factors for implementing gender equality interventions in R&I: Structures and processes. *Evaluation and Program Planning 77.* doi: 10.1016/j.evalprogplan.2019.101726

Pattinson, Steven, Preece, David, & Dawson, Patrick (2016). In search of innovative capabilities of communities of practice: A systematic review and typology for future research. *Management Learning*, S. 1–19. doi: 10.1177/1350507616646698

Pawlowsky, Peter, & Geppert, Mike (2005). Organisationales Lernen. In Elke Weik, & Rainhart Lang (Eds.), *Moderne Organisationstheorien 1. Handlungsorientierte Ansätze* (S. 259–294). Gabler. doi: 10.1007/978-3-322-90466-9

Pyrko, Igor, Dörfler, Viktor, & Eden, Colin (2017). Thinking together: What makes Communities of Practice work? *Human Relations 70(4)*, S. 389–409. doi: 10.1177/0018726716661040

Roberts, Joanne (2006). Limits to Communities of Practice. *Journal of Management Studies*, 43, 623–639. https://doi.org/10.1111/j.1467-6486.2006.00618.x

Thomson, Aleksandra, Palmén, Rachel, Reidl, Sybille, Barnard, Sarah, Beranek, Sarah, Dainty, Andrew R. J & Hassan, Tarek M. (2021). Fostering collaborative approaches to gender equality interventions in higher education and research: the case of transnational and multi-institutional communities of practice. *Journal of Gender Studies*. https://doi.org/10.1080/09589236.2021.1935804

Verge, Tania, Ferrer-Fons, Mariona, & González, M. Josè (2018). Resistance to mainstreaming gender into the higher education curriculum. *European Journal of Women's Studies 25(1)*, 86–101. doi: 10.1177/1350506816688237

Vincent, Katharine, Steynor, Anna, Waagsaether, Katinka, & Cull, Tracy (2018). Communities of practice: One size does not fit all. *Climate Services 11*, S. 72–77. doi: 10.1016/j.cliser.2018.05.004

Wenger, Etienne, McDermott, Richard A., & Snyder, William (2002). *Cultivating Communities of Practice: A Guide to Managing Knowledge*. Harvard Business Press.

Wenger, Etienne, Trayner, Beverly, de Laat, Maarten (2011). Promoting and assessing value creation in communities and networks: A conceptual framework. Ruud de Moor Centrum https://www.researchgate.net/publication/220040553_Promoting_and_Assessing_Value_Creation_in_Communities_and_Networks_A_Conceptual_Framework

Wroblewski, Angela (2021). *GENDER ACTION D3.2 Monitoring of ERA priority 4implementation*.https://genderaction.eu/wp-content/uploads/2020/03/D3.2._MonitoringERApriority4implementation.pdf.

Zboralski, Katja, Salomo, Soeren, & Gemuenden, Hans Georg (2007). Organizational Benefits of Communities of Practice: A Two-Stage Information Processing Model. *Cybernetics and Systems 37(6)*, S. 533–552. doi: 10.1080/01969720600734461

11 Scaling up

From Communities of Practice to
the three ERA gender equality and
mainstreaming objectives: careers,
decision-making and integrating
the gender dimension into research
and educational content

*Kathrin Rabsch, Rachel Palmén, Maria Caprile,
Claartje Vinkenburg and Karolina Kublickiene*

Introduction

Whilst other chapters in this volume have been developed to reflect on the experiences of the different Communities of Practice (CoPs) supported by the ACT project (see Chapters 5, 6, 7, 8, and 9), this chapter aims to highlight how knowledge sharing between and beyond the CoPs has been achieved specifically in relation to the three European Research Area (ERA) objectives for the gender equality and mainstreaming priority such as careers, decision-making, and the gender dimension in teaching and research content. In *scaling up* the CoP approach, ACT established so-called ERA priority coordination groups or ERAGs, in which cross-cutting issues related to each of the three priorities were identified and addressed. These groups brought together ACT consortium core and seed partners, members of different CoPs, ACT advisory board members, experts, representatives of ERA level players and other relevant research and innovation (R&I) representatives from the CoPs' contexts – local, regional, national, and disciplinary networks. Scaling up the CoP approach does not necessarily refer to enlarging CoPs in terms of size or geographic reach, but rather to increase and strengthen the scope of their work towards creating a sustainable impact at national and European level by sharing knowledge and experience between and beyond CoPs from a range of disciplinary fields and geographical contexts.

We aligned scaling up with the three ERA gender equality and mainstreaming objectives targeting careers, decision-making, and the integration of the gender dimension in research and educational content (see Chapter 1

DOI: 10.4324/9781003225546-11

for a discussion of the policy context of these three objectives). Through involving different ERA level players (such as LERU, EURODOC, EUA etc.), the CoPs connected with relevant stakeholders, shared their experiences and learnings and promoted the CoP approach at the ERA level through dialogue with those stakeholders. The ERA priority coordination groups provided a forum for discussing ongoing activities *between* CoPs whilst coordinating insights and progress made in each thematic area effectively aiming to scale-up to the ERA level. Due to the fact that ACT's practices were largely bottom-up, demand-sensitive, and validated in practice, we aimed for a scalable process for integrating ERA priorities in new and/ or existing R&I plans and institutions. This chapter therefore reflects on the experience of these three ERA coordination groups and subsequently organised events. Four of the authors coordinated these three groups whilst the other author had the overall responsibility for coordinating the Matching Events. This chapter is based on reflections of the authors, the e-discussion records as well as the subsequent reports and minutes taken from the Matching Events. First however, we briefly describe some of the more common interventions in each of the three areas in order to ground the chapter within the current landscape of interventions promoting gender equality in R&I throughout Europe.

Careers

Firstly, following from the prioritisation of careers as an important topic for gender equality policies, HEIs, research performing and research funding organisations, national governments and professional associations have made provisions for career-related issues in their formalised gender equality plans and policies. A comprehensive but not exhaustive list of such efforts includes:

- Efforts to promote research careers among girls and young women, especially STEM fields;
- Efforts to improve career advancement, representation, and retention of women in research careers, especially where men are currently overrepresented;
- Efforts to promote work-life balance;
- Efforts to reduce precariousness (e.g., temporary contracts), especially where women are overrepresented in such positions;
- Efforts to transform institutional culture, promote inclusiveness, and prevent sexual harassment.

Decision making

Efforts for achieving gender balance in decision making bodies and foster gender competence have been developed in different ways. The report on the implementation of targets by the ERAC SWG GRI highlights that several

Member States and Associated Countries have made progress and are developing their national as well as institutional policy frameworks to advance gender balance in decision-making. The analysis of ERA National Action Plan implementation shows that 27 actions on the national level (14.6% of the total number of actions related to gender equality) focus on decision making. They mainly deal with:

- Setting guiding targets or quotas through legislation
- Funding/mentoring/support programmes for women professors
- Developing Gender Equality Plans (GEPs) or similar initiatives
- Training and guidelines to raise gender awareness and counteract gender bias
- Introducing incentives for institutions adopting pro-active measures, and/or sanctions for non-compliance, as necessary (ERAC, 2020).

Some of these actions are stand-alone initiatives while others are part of a comprehensive policy where addressing gender balance in decision-making is one part. As well as the above, institutional level initiatives may also include developing election rules or developing gender competences for leadership.

Gender dimension in the research and educational content

Of the three objectives, integrating the gender dimension in research and educational content, as well as in innovation content[1] is arguably the objective where the least progress has been made (European Commission, 2017). Various initiatives have been taken by funding agencies, universities and peer reviewed journals in an attempt to promote the integration of the gender dimension into research and educational content. Activities to integrate the gender dimension in educational content can include:

- Developing new knowledge and training methods for students at different levels and researchers in fields where sex and gender analysis is of special relevance (e.g., Karolinska Institute in health and biomedical research)
- Including methods of sex and gender analysis and related knowledge in curricula (GENERA, 2019; LERU, 2015)

The inclusion of the gender dimension throughout all stages of the research cycle (idea/proposal/research/dissemination) can be promoted. Research Performing and Funding Organisations are encouraged and implementing the tasks to:

- Develop specific funding criteria to mainstream sex and gender analysis in R&I content and programmes (as in Horizon 2020/Horizon Europe)

- Ask research applicants to address 'how sex and gender analysis is taken into account in the project's content' (Horizon 2020; Science Foundation Ireland)
- Raise gender awareness and competence for applicants, reviewers or evaluation panels, providing specific guidance and training (LERU, 2015)
- Support gender-related fields of research (Horizon 2020)
- Provide tools for researchers to understand and apply perspectives of sex and gender in research and educational content, methods in their research fields, for instance via training workshops, seminars, under-graduate and postgraduate courses or showcasing good examples (European Commission, 2020; GENERA, 2019; LERU, 2015)
- Include training in sex and gender analysis as eligible costs in applications (Science Foundation Ireland, 2016).

Peer reviewed journals have also employed guidelines to apply sex and gender perspectives as one criterion among many when evaluating and selecting manuscripts for publication (Schiebinger et al, 2011–2020).

The ERAGs and subsequent Matching Events provided a forum whereby experiences of the different initiatives currently being developed to advance these three objectives by various stakeholders including the ACT CoPs, as well as representatives from research funding and performing organisations will be shared.

Methodology of Matching Events and scaling up

The ACT Matching Events were held virtually throughout the month of October 2020 – a total of 198 people attended the 7 sessions. The events were supposed to take place face-to-face, but due to the COVID-19 pandemic, they were held on-line. The events had three main aims: to facilitate exchange amongst the CoPs; to enable the scaling up of the CoP work to the national and international (policy) levels; as well as reaching out to previously isolated yet interested parties.

The main aim of the Matching Events was to facilitate exchange between the different CoPs – to share experiences, best practices and knowledge on a specific topic. CoPs' activities, experiences and knowledge development were exchanged through providing a platform where CoPs could come together with experts, Advisory Board members and other interested parties and stakeholders. Scaling up refers to the promotion and scaling up of the work of the ACT CoPs, to the national and international levels by inviting different ERA level players (EURODOC, LERU, EUA, EWORA, etc.), policymakers and other national and international stakeholders. Connecting with strategic stakeholders was seen to further develop the CoPs' work as well as scaling up their impact. Reaching out refers specifically to including those institutions that are to date isolated from gender equality in R&I

networks throughout Europe yet, however have an interest and motivation in learning and connecting. A key aspect of 'Matching' was, therefore, to invite organisations and people that expressed interest in the ACT project and its CoPs yet belonged to organisations and regions that have been identified as 'structural holes' in the ACT Community Mapping survey (Reidl et al, 2019). These are regions and organisations who are not yet connected to the network of European projects addressing gender equality, even though they would like to become part of the said network. Reaching out and expanding inclusivity may, therefore, result in more organisations getting involved, enlarging the ACT network and connecting CoPs with potential new members.

The methodology of the Matching Events included designing interactive and collaborative events that allowed for and fostered the exchange of knowledge, sharing good practices and experiences. The use of Mural and the breakout sessions added value to these events as they supported co-creation, working collaboratively and enabled a feeling of togetherness even if the events took place virtually. In the breakout sessions, representatives from different organisations and/or CoPs had the opportunity to discuss different topics while reflecting on their own perspectives, needs, and backgrounds. The Matching Events were a series of events with seven sessions in total. One session addressed the GEP requirement as an eligibility criterion for Horizon Europe funding. Two sessions each were dedicated to careers, decision-making, and the gender dimension.

Gender Equality Plan eligibility criterion Horizon Europe

The first session of the Matching Events discussed the GEP eligibility criterion for Horizon Europe. This event counted on 63 participants from over 20 countries. An EC representative presented the EU Policy on Gender Equality in Research & Innovation and stressed the new eligibility criterion and its implications, framework and conditions. The most prominent aspects presented were the new Gender Equality Strategy (2020–2025), with new measures to strengthen gender equality in Horizon Europe, a review of past Gender Equality Projects as well as the structure of Horizon Europe. Beyond the eligibility criterion that requires GEPs from applicants (applies to public bodies, research organisations and higher education institutions), Horizon Europe supports the advancement of gender equality by including the integration of the gender dimension in the R&I content and strengthens the gender dimension across the programme. A fruitful and interesting discussion was held which included the following points:

- How this eligibility criterion will be implemented in regard to assessing the organisation's GEP.
- How organisations will be supported prior to, as well as during the one-year transition period (e.g., by national contact points or rather on an

organisational level). At the time of the discussion – the exact design of the support was yet to be defined. An important point mentioned by participants was the need to provide expertise at both the national level and at the institutional level. The EC plans to provide support in the form of a knowledge and support facility to prepare for the implementation of the GEP criterion.

- How can it be ensured that the criterion will not just become a 'box-ticking' exercise? How will the EC evaluate the GEPs? Will there be a follow-up? Rather than assessing the GEPs that organisations declare they have in place, emphasis will be placed on regular checks throughout the course of the Horizon Europe programme as to whether organisations are addressing all relevant aspects. Organisations furthermore need to have their GEPs publicly available on a webpage and they need to be endorsed by the highest authority of the organisation.
- The GEP eligibility criterion by the EC was welcomed by participants. Nevertheless, some reservations were also articulated with regards to the requirement becoming a 'box-ticking' exercise and whether this approach is sufficient to change the system.

This event provided the broader policy context for the following three Matching Events, presented in the following sections, which focused on careers, decision-making and gender dimension in research and educational content towards innovations.

Careers

For the careers priority, the ERA coordination group made an informal inventory of urgent topics to discuss cross-cutting CoPs at the beginning of 2020. We held an e-discussion[2] on the prevalence and prevention of sexual harassment in research organisations.

The careers Matching Events were held on the 27th and 29th of October 2020. The session on the first day focused on COVID-19 and gendered career consequences. The following issues were discussed: (1) What exactly are career consequences resulting from COVID-19? (2) How can these consequences be measured? (3) What are the compensating measures for researchers? (4) What is the state of the art for work-family measures? The second session was dedicated to the DORA[3] declaration and evaluating researchers/academics. Both sessions included group work.

During the Matching Events, we zoomed in on COVID-19 and its disproportionate effects on women in research and academia. Evidence was presented of the possible career consequences of COVID-19 –indicating that cumulated disadvantages disproportionately affect women academics for which we need to find creative solutions. The impacts of COVID-19 have been experienced at various different levels, mental distress, delays in research, difficulties regarding multiple responsibilities of parents as well as work insecurity at an individual level. At an organisational level – the impact of COVID-19 is evident – troubles managing a team during a global crisis or delays in providing equipment, as well as keeping up international collaborations were identified as major challenges to be dealt with during the pandemic. Research has shown that gender imbalances at home intensify the difficulties for women, for example resulting in less academic activity. GEinCEE CoP representatives presented the preliminary findings of the research they have undertaken on measuring these consequences- throughout four key dimensions, academic work impact, psychological strain, work-life balance and support. These findings demonstrated that women declared more often (than men) that COVID-19 has had a negative impact on their research, academic activities, and promotion.

When considering the gendered career consequences of COVID-19 and their measurement – three different compensation measures for researchers when applying for research funding were presented. Firstly, 'Extension', a fairly common instrument, with the risk however of a backlog of activities. Secondly, 'Supplementation', which refers to (mostly) monetary support. Thirdly, 'Adaptation of Criteria' -which is rather rare and controversial. The assumption is that there are certain expectations towards researchers, which will be lowered because of care responsibilities. Australian universities have an approach called 'Achievement Relative to Opportunity (AR20) – which takes only productive months into account.

Recent research documenting the state of the art of work-family measures – including maternity leave was presented (LERU, 2020). The report deals with why family leave is so problematic – citing four main reasons: 1) different attitudes towards men and women as carers, 2) the management of family leave, 3) impact on research and 4) impact on individual careers. How this can now be linked to the COVID-19 situation? Whilst there are no clear answers – perhaps the pandemic may bring opportunities – as men are currently experiencing the double burden (more than ever) – which might lead to change.

From the discussions held in the e-discussion and the matching events, we identified various sources of innovation potential for the careers priority. We discussed the potential and the perils involved in the adoption of so-called narrative CVs in research assessment (selection, promotion, funding). While narrative CVs help us move away from counting publications and citations towards a more qualitative assessment of impact, the evidence of the cumulative careers effects of gender, race, and class bias as evidenced in the language of evaluation points to problems in simply adopting this approach (Vinkenburg et al, 2021). Research funders (united in FORGEN and the DORA funders group) are joining forces in evaluating the (un)intended consequences of the narrative turn (see Fritch et al, 2021). In a related vein (as also relevant to our assessment of researchers), we talked about the norm of individualised hyper-competition and the superstar model that is still prominent but increasingly obsolete given the prominence of collaborative teamwork in many disciplines. Simple solutions such as emphasising and valuing teamwork and collaborative skills in job descriptions and call materials help to bend the norm. Finally, we discussed ways to measure career advancement beyond mere representation as is commonly done in the scissor graphs and about operationalisations of intersectionality in data collection and monitoring.

Decision-making

For the decision making ERAG – we first held an e-discussion[4] on gender equality in decision making in HE and R&I where we invited key experts to engage with participants to discuss some of the key debates in this field including quotas and targets, gender competence in decision-making bodies and power followed by a more practical focus on strategies for action. This was followed by the Matching Event on decision-making which was comprised of two sessions held on the 14th and 16th of October 2020.

The first session was dedicated to CoPs and 'Collaborative Action: Gender Equality in Decision-Making in R&I', where three ACT CoPs (GenBUDGET, FORGEN and LAC CoP) shared insights on their work. The GenBUDGET CoP presented their work which includes developing best practices on gender impact assessment, of financial and management mechanisms at research organisations as well as attaining deep knowledge on the gendered impact and outcomes of the financial and management mechanisms at research organisations. One of their main aims however is to deepen the knowledge and expertise on how to implement gender budgeting in research organisations (see Chapter 8). The FORGEN CoP presented the work that they have carried out mitigating gender-bias in decision-making in evaluation of research projects. Their work has included mapping gender equality measures in the grant evaluation process. The first step was to define the steps and processes of the grant evaluation process in the different funding agencies to form one grant evaluation process flow diagram. Minor differences in the grant evaluation process can have a large effect bias within the process. Learning and sharing knowledge on the different types of grant evaluation processes and their implementation was recognised as key to understanding how to mitigate bias within the process. The Latin American CoP presented their work on gender equality in decision-making – specifically for institutional change. This was followed by an example for best practice, the AKKA programme – a gender integrated leadership programme at Lund university. By changing the structures and cultures, the AKKA program increased the proportion of women in leading positions as well as the visibility of women as potential leaders – whilst the gender awareness of leaders was raised. Key findings of this session included:

- It is important to include executive boards and deans to have their support and to introduce the work of the CoPs but also the other way around, that CoPs provide support and inspiration to decision-making bodies.

- It is important to include actors from different levels (e.g., students) to put pressure on decision-makers on the higher levels.
- Collaboration beyond the institution is one of the key aspects for CoP progress.
- CoP members not only benefit from learning from others and co-developing new knowledge and practice. Support from outside the institution entails increased leverage to act within the institution.
- Cooperation between CoPs focusing on cross-cutting themes of mutual relevance enlarges potential benefits and impact.

The second session focused on 'National Policy and Institutional Measures for Gender Balance in Decision-Making and Leadership Positions in R&I', with different presentations on examples for national or institutional measures and an overview of national legislative and policy measures as well as group work. The Chair of the Standing Working Group on Gender in Research and Innovation, also co-author of the Report on the Implementation of Targets: Follow-up on the 2018 Guidance Recommendations – presented the main conclusions of this report. Findings assessed whether and to what extent the Member States and Associated Countries have adopted the seven recommendations launched in 2018:

- Collect and publish sex-disaggregated data on the composition of professorship and management/leadership positions.
- Promote gender balance in decision-making positions and professorships with adequate awareness raising and training.
- Institutionalise gender equality plans as an assessment tool in the accreditation of universities and make them mandatory for universities and research organisations.
- Institutionalise the proportion of women in grade A/professor positions as an assessment criterion in institutional evaluations (higher education accreditation, performance contracts with universities).
- Set and implement guiding targets and/or quotas through legislation.
- Evaluate regularly the implementation of quotas and/or targets.
- Introduce incentives for institutions adopting pro-active measures and/or sanctions for non-compliance, as necessary. (European Commission, 2018, p. 4).

The results show that some recommendations have been implemented by almost every country, whereas other recommendations have only been implemented by a few. The recommendation to collect and publish sex-disaggregated data on the composition of professorship and management/leadership positions for example has been implemented by 92% of the countries. In contrast, the recommendation to institutionalise the proportion of women in grade A/professor positions as an assessment criterion on institutional evaluations has only been implemented by 16%. A main topic for discussion was the extent to which the percentage of women in Grade A is a useful indicator for gender equality in R&I – given that this indicator tends to be strongly influenced by contextual factors (i.e., the weight of the business and enterprise sector in R&I). The Irish case study was then presented whereby gender balance in decision making forms part of a comprehensive National Gender Equality Policy in HE. Political backing was identified as a key factor for the successful implementation of such an approach. Another approach taken at the institutional level – was presented – the change in election rules for its board of governors to ensure a 40/60 gender balance implemented by Ghent University in Belgium.

A central idea shared by all attendants was the need to go beyond the 'numbers': achieving gender balance in decision making is necessary, but not enough. Decision-making has to be gender competent, and this involves all people in decision-making positions. In this sense, the concept of leadership and a narrow view of gender balance has to be problematised and the building of gender competence in decision-making bodies (irrelevant of the sex/gender of membership of these bodies) – seen as a key issue. A second issue was the importance of gender budgeting. Gender budgeting as conceptualised and implemented by one of the ACT CoPs is a powerful approach to foster evidence-based gender-sensitive decision-making. Participants also agreed that there is not one generally applicable approach and that it is important to tailor relevant measures and actions to the institutional and wider contexts. This applies to all kinds of measures, including different rules or measures to foster gender balance.

Concerning decision-making, the debate held during the e-discussion and the Matching Events was fruitful in terms of identifying challenges and potential innovations. A first aspect was the need to develop different approaches to implement targets or quotas building on institutional key stakeholders' support. Even when quotas are mandatory according to legislation, institutional support is important. Institutions have the ability to balance decision-making bodies through institutional rules and specific measures. A second aspect is expanding the implementation of training on gender bias and gender competence in decision-making bodies. This includes initiatives aimed at promoting a more inclusive type of leadership. Finally, expanding the use of gender budgeting and fostering gender-aware decisions (also linked to gender in research content and innovation) were considered key strategies.

Gender dimension in research content and teaching

On the 20th and 22nd of October 2020, the Matching Events addressing the gender dimension in research and educational content to foster the potential of innovations were held. These took place in conjunction and collaboration with other research educational activities at the Karolinska Institutet, which also included participants and stakeholders involved in the GENDER-NET grant application with Canadian partners. Moreover, the participants of the Matching Events had the possibility to participate in additional educational activities organised by Karolinska Institutet. The two main sessions tackled the following three topics:

- Why does integrating the gender dimension add value to research and education in the field of STEM?
- Meet an expert get inspired: from one size fits all to an individualised approach.
- When and how? Hands on with practical work. Becoming a gender champion to boost your science and education towards innovation.

The first session focused on integrating the gender dimension from a European perspective. Integrating the gender dimension is needed in order to ensure reproducibility, excellence, and societal value – but it is also necessary to research what it means when sex differences are ignored, for example, the unexpected side effects of drugs. Another discussion point included how change processes can work – a multifaceted approach operating on different levels was seen as important through building knowledge, ensuring a critical mass, working at different structural levels, engaging leadership, and putting pressure on the organisation from the outside.

The most advanced and straight forward examples provided were the reflections of sex and gender in medical clinical health care practice and translational research that could have an impact towards precision medicine. The importance of sex and gender perspectives are increasingly appreciated in the biomedical field and, therefore, are crucial for precision medicine. Moreover, the existing discussions and further arguments strengthen the suggestion that those aspects are not only important for researchers, but also important to engage other actors such as funding agencies, leaders of universities and peer reviewed journals. Integrating the gender dimension into research from different disciplines including health sciences, with a specific focus on 'Sex Differences in Alzheimer's Disease' and in the field of ICT with a specific focus on Artificial Intelligence was presented and discussed by invited speakers. In the case of preclinical research, if animal experiments are performed on male mice or sex unidentified and non-reported animals and/or cell cultures treatments, preventative recommendations and the diagnostic process are most likely to be undifferentiated between men and women, and therefore adding the risk of adverse outcomes or failure of appropriate treatment, prophylactics, and/or diagnostic tools. This highlighted the need for a more integrative scientific approach in this field by means of developing specific research projects, performing sex disaggregated statistics and/or appreciating sociocultural/gender sensitive variables when feasible.

In the break-out sessions, participants discussed the following issues in separate groups:

- Which disciplines and topics are of importance, need further implementation and/or strengthening of the gender dimension and should be prioritised, as well as which terms, methods, approaches would be best to promote their understandings and implementation e.g., at different levels of educations and/or research processes (e.g., sex and gender per se, sex disaggregated statistics, gender sensitive variables);
- How gender bias free data generation and knowledge dissemination can be ensured (e.g., requirements from publishers, follow-up from funding organisations how the gender dimension was implemented in research progress and reporting);
- From the perspective of using the CoP approach – what are the most important action points to boost the implementation of integrating the gender dimension when applying a top-down approach.

At this stage, several action points and recommendations were identified in the discussion:

- To develop/have guidelines on how to use existing data and how to identify potential bias/weaknesses
- To make proposals which are sex and gender sensitive
- Promoting the idea of including gender dimension in education
- Getting the leadership on board by raising awareness of the need to integrate the gender dimension in research and educational content
- How to add gender sensitive variables to the technology to make precise assessments and developments that could serve everybody's needs

In addition, the participants had the opportunity to share the knowledge by the means of available tools and knowledge data bases. Those developed by collaborative efforts of the European Union, Stanford University in the USA, Karolinska Institutet, and even the Canadian Institute of gender and health. This also included reflections on how the ACT's GENERA CoP could serve as an optimal example of how to implement the gender dimension successfully in selected fields of STEM. Furthermore, new alliances within new disciplines emerged, as well as a new CoP in Israel, NOGAFEM[5] with a focus on implementing the gender dimension to optimise women's health, leadership and innovations.

Innovation potential and main conclusions

To meet the scaling up goals, the Matching Events held for each of the ERA priorities provided an opportunity for members of the different CoPs to meet and exchange experiences, good practices, and lessons learnt. Participants highlighted the value and benefits of the exchange – particularly empha-sising the desire to continue this exchange amongst the CoPs beyond the boundaries of their own CoP including those at different developmental stages. While CoPs provide support from the outside of the institution to work at the institutional level, cooperation among different CoPs turned out to be especially important for supporting collaboration on common chal-lenges – regardless of the disciplinary field or geographical context.

An important aspect is that collaboration involved not only CoP mem-bers, but also different stakeholders that participated in the Matching Events – ERA level players such as representatives of EURODOC, LERU, EUA, EWORA, etc. This meant that the work of these networks could be presented to the ACT CoPs – whilst the work being carried out by the CoPs were presented to these ERA level players – aiming to forge future synergies as well as scaling up the impact of work carried out.

Furthermore, the Matching Events also enabled us to reach out beyond existing networks. Through using the results of the community mapping – we specifically targeted those regions that to date are not connected to gen-der and science networks throughout Europe – yet are interested in taking part. By addressing these structural holes, inclusivity was expanded as well as the ACT network and its CoPs.

These three aspects highlight how scaling up the CoP approach by focus-ing on cross-cutting issues has an enormous potential for increasing impact on a European and even global scale. In this sense, the FORGEN CoP can be seen as a good example of scaling up. FORGEN started to collaborate with the DORA Funders Group, working together on how to mitigate bias when assessing applicants as explained in the FORGEN blog[6]. This collab-oration culminated in joint workshops organised jointly by the FORGEN CoP and DORA, on using Narrative CVs: process optimisation and bias mitigation. The focus of the initial workshop was on how funders can align and optimise the process for assessing narrative CVs, as well as discussing the best ways to improve the evidence-base for iterative policy improve-ment. Over 120 participants attended from over 50 research and innovation funding agencies. A follow-up workshop is planned for early 2022 to focus on defining key principles for the design, implementation, and assessment of narrative CVs.

The work carried out to scale up the CoPs focusing on the three ERA pri-orities has shown the potential of further sharing and co-developing knowl-edge and practice to align efforts and increase impact at the institutional, national, and European level.

Notes

1. A gender dimension in the context of Horizon Europe refers to the integration of sex/gender analysis and methods in the research content, where 'Sex' refers to the biological characteristics of beings, whether female, male, or intersex,

involving different levels of expression for genes, gametes, organ and systems morphology and function beyond their reproductive functions. 'Gender' refers to socio-cultural processes that shape behaviours, preferences, values, products, technologies, knowledges, and how individuals and groups interact with their environment. Importantly, the two terms may interact with and influence each other (Schiebinger, Londa et al.) Gender http://genderedinnovations. stanford.edu/terms/gender.html (2011–2019)).
2. See https://www.genderportal.eu/group/e-discussion-addressing-sexual-harassment-research-organizations.
3. See https://sfdora.org/.
4. See https://www.genderportal.eu/group/e-discussion-gender-equality-decision-making-ri-and-he.
5. See https://www.nogafem.com/.
6. See https://forgen.act-on-gender.eu/Blog/summary-forgen-cop-outputs-and-events-date.

References

Council of the European Union. (2015). *Advancing gender equality in the European Research Area – Council conclusions (adopted on 01/12/2015)* (No. 14846/15). Council of the European Union. https://data.consilium.europa.eu/doc/document/ST-14846-2015-INIT/en/pdf

Standing Working Group on Gender in Research and Innovation (ERAC). (2020). *Report on the Implementation of Targets: Follow-Up on the 2018 guidance recommendations* (WK 8491/2020 INIT). European Research Area and Innovation Committee. https://genderaction.eu/wp-content/uploads/2020/08/Report-on-the-Implementation-of-Targets.-Follow-Up-on-the-2018-Guidance_ERAC_SWGRRI. pdf

Standing Working Group on Gender in Research and Innovation (ERAC). (2018). *Report on the implementation of Council Conclusions of 1 December 2015 on advancing gender equality in the European Research Area* (No. 1213/18). European Research Area and Innovation Committee. https://data.consilium.europa.eu/doc/document/ST-1213-2018-INIT/en/pdf

European Commission, Directorate-General for Research and Innovation. (2017). *Interim evaluation: Gender equality as a crosscutting issue in Horizon 2020.* Publications Office of the European Union. https://data.europa.eu/doi/10.2777/054612

European Commission, Directorate-General for Research and Innovation. (2018). *Guidance to facilitate the implementation of targets to promote gender equality in research and innovation*, Publications Office of the European Union. https://data.europa.eu/doi/10.2777/956389

European Commission. (2019). *ERA progress report 2018: The European Research Area: Advancing together the Europe of research and innovation.* Publications Office of the European Union. https://data.europa.eu/doi/10.2777/118067

European Commission. (2020). *Gendered innovations 2: How inclusive analysis contributes to research and innovation: Policy review.* Publications Office of the European Union. https://data.europa.eu/doi/10.2777/316197

Fritch, Rochelle, Hatch, Anna, Hazlett, Haley, & Vinkenburg, Claartje. (2021). Using Narrative CVs: Process Optimization and bias mitigation, https://zenodo.org/record/5799414#.YeA_yljMKfU

Gender equality network in the European research area (GENERA). (2019). Accessed 27 January https://genera-project.com/

League of European Research Universities (LERU), & Hopkins, A. (2020). *Family leave for researchers at LERU Universities* (No. 28; Advice Paper). League of European Research Universities. https://www.leru.org/files/Publications/LERU-Family-Leave-Paper-Final.pdf

League of European Research Universities (LERU). (2015). *Gendered research and innovation: integrating sex and gender into the research process.* (No. 18; Advice Paper). League of European Research Universities. https://www.leru.org/files/Gendered-Research-and-Innovation-Full-paper.pdf

Reidl, Sybille, Krzaklewska, Ewa, Schön, Lisa, & Warat, Marta. (2019). ACT community mapping report: cooperation, barriers and progress in advancing gender equality in research organisations. *Zenodo.* https://doi.org/10.5281/zenodo.3247432

Schiebinger, Londa, Klinge, Ineke, Paik, Hee Young, Sánchez de Madariaga, Inés, Schraudner, Martina, and Stefanick, Marcia. (Eds.) (2011–2020). Gendered Innovations in Science, Health & Medicine, Engineering, and Environment (genderedinnovations.stanford.edu).

Science Foundation Ireland. (2016). *Gender Strategy 2016–2020.* Science Foundation Ireland. https://www.sfi.ie/resources/SFI-Gender-Strategy-2016-2020.pdf

Vinkenburg, Claartje J., Ossenkop, Carolin, & Schiffbaenker, Helene. (2021). Selling science: Optimizing the research funding evaluation and decision process. *Equality, Diversity and Inclusion: An International Journal.* https://doi.org/10.1108/EDI-01-2021-0028

Wenger-Trayner, E., & Wenger-Trayner, B. (2015). *Introduction to Communities of Practice.* https://wenger-trayner.com/introduction-to-communities-of-practice/

12 Concluding remarks

Towards Communities of Practice for greater gender equality

Rachel Palmén and Jörg Müller

As set out in the introductory chapter, this book provides an explicit account of Communities of Practice (CoPs) as an instrument for accelerating gender equality and institutional change in research & innovation (R&I) and higher education (HE) across Europe and beyond. It is grounded in our experiences of setting up, supporting and facilitating eight CoPs of over 144 organisations working together to further strengthen gender equality during ACT. This was a three and a half year project financed by the European Commission under the Horizon 2020 framework program. Contributors to this edited volume have provided their reflections on this experience as CoP facilitators, gender equality practitioners, researchers, gender experts, scholars and evaluators. Despite coming from various disciplinary backgrounds, the authors have come together to act on gender equality and push forward institutional change in R&I and HE through engaging in inter-organisational cooperation. This book has charted these experiences.

In this concluding chapter, we revisit one of the main tensions that defined the ACT project – namely how can institutional change be supported by inter-organisational collaboration as described in the introduction. How this has been achieved is explored through the three conceptual dimensions that constitute a CoP: 'domain', 'community', and 'practice'. Furthermore, we aim to weave together the more theoretical chapters with the grounded experiences of the diverse CoPs that have been set up and supported. This will also give us the opportunity to revisit some of the main issues and key questions highlighted in the introduction and offer our reflections built on the grounded experiences of the eight CoPs. How can locally generated solutions to problems be re-inserted and aligned with overarching goals? To what extent have the CoPs provided a 'safe space' to establish collaborative relations, yet how effective can this be in a hyper-competitive neo-liberal context with ever increasing pressures? To what degree can CoPs activate a more collective and political mode of action? Can a CoP approach be harnessed to redistribute resources and privileges – as needed in gender equality work? How can CoP practices be cultivated through the inter-organisational CoP – but become embedded with a particular institution? These are some of the questions to which we want to return in this concluding chapter.

DOI: 10.4324/9781003225546-12

Domain and vision

Whilst initiatives to further gender equality in R&I and HE can be seen as the overall domain – defined by the three European Research Area (ERA) objectives regarding 'careers', 'decision-making' and integrating the 'gender dimension' in research and educational content – CoPs furthermore developed and elaborated the details of their shared interest in a self-determined manner. Although CoPs can emerge bottom-up or be setup on purpose using a top-down approach (Wenger et al., 2002), articulating a shared 'vision' among its members was deemed crucial in the context of the ACT project. First, because Gender Equality Plans (GEPs) or ERA gender equality objectives appear as too broad as to be really useful for a CoP agenda. As Reidl, Beranek & Holzinger (p. 170 this volume) comment 'CoPs are not limited to the topic of GEPs, since not all institutions might be in the right time and place (i.e., context) to tackle such a complex undertaking'. Along these lines, the Slovenian CoP for example decided to call itself Alt + G (Alternative Gender Infrastructure) as an explicit recognition that there is a need to establish other infrastructures than GEPs to encourage the take up of gender equality measures and initiatives in their specific regional, post-Yugoslavian context.

Second, leaving room for each CoP to co-create their vision and define their work area was key in garnering the support and buy in from CoP members. Seed partners initially drafted their 'vision' of the CoP – this was often linked to their expertise, existing networks, past and present involvement in structural change projects, and desire to work on a specific area. Potential CoP members were then invited to further contribute to this vision. This initial expertise, as well as the discussions held with potential CoP members ensured that CoPs responded to and helped to define the real needs of their members. This has to be key, in a context of a neo-liberal academy where so many demands compete for precious time and CoP members work largely on a voluntary basis. If CoP members are motivated to participate and engage – they must see the benefits of their engagement. As Thomson, Rabsch, Barnard, Hassan & Dainty comment in Chapter 2, the social and design choices made when setting up CoPs must be purposeful and contextually relevant. They go on to highlight how 'communities that succeed and that last, are characterised by focused and well-defined purposes that are linked to the strategic mission of the sponsoring organisation. The most effective way to define a CoP's purpose is to assess how this initiative will benefit the community's stakeholders and what specific needs are to be met by the community' (see Chapter 4, p. 68).

The specific domain of each CoP was developed in response to the real needs of the seed partners and recruited CoP members. This resulted in the setting up of two geographically bounded CoPs, one in Central and Eastern Europe and one in Slovenia (GEinCEE and Alt+G), two thematically based CoPs, one implementing gender budgeting (GenBUDGET) and one examining sustainability (STRATEGIES) and two disciplinary CoPs, one in Life Sciences (LifeSciCoP) and one in Physics (GENERA). As mentioned, in addition a CoP for Research Funding Organisations (FORGEN) was created as well as a CoP of Latin American Universities (LAC CoP). Even though the experiences of these latter two CoPs are not presented as an individual chapter in this book, they have nevertheless contributed to the overall discussion and results either via the evaluation or simply throughout the various project activities. The diversity in terms of geographical scope, organisation type, or disciplinary orientation of CoPs has provided the ACT project with a rich landscape of approaches towards the setup and management of CoPs as well as the creation of new knowledge and practice for greater gender equality.

Geographical-based Communities of Practice

Two geographically based CoPs were set up in Europe – one covering Central Eastern Europe (GEinCEE) and one operating in Slovenia (Alt+G). In their chapter, Reidl, Baranek & Holzinger (p. 168 this volume) highlight the regional and geographical disparities across Europe as to the extent to which gender equality in R&I is a national priority area for action. They draw our attention to the widening gap between countries that are 'experienced' and those that are 'inactive' regarding gender equality in R&I at a national policy level. There is a valid preoccupation with the best ways to tackle these increased inequalities in terms of policy action/inaction. Whilst regionally and contextually targeted solutions may make sense, it seems that critics fear that more 'inactive' areas may experience 'ghettoisation' and become even more excluded from the debate. However, as the experience of setting up two regionally based CoPs has demonstrated, the opportunity to concentrate on the needs and particular idiosyncrasies of a specific regional/national environment was seen to be key for the CoPs' success. As Sekula, Ciaputa, Warat, Krzaklewska, Beranek & Reidl (p. 87 this volume) note, 'the discussions with the members-to-be and the results of the Community Mapping reinforced the belief that the regional focus is important as countries within Central and Eastern Europe (CEE) share similar aims, concerns, needs and institutional context. Indeed, the regional focus of the CoP is seen as its main strength, as underlined by the members. Focusing on the CEE region is beneficial in terms of knowledge sharing and providing a sense of belonging: "I think it is important that this community of practice is focused mainly on this region" (Member 3 interview)'.

The geographical focus that spanned various Member States was seen to be useful, not only facilitating a sense of community and shared identity but also for generating new knowledge that is especially pertinent to the CoP members' needs. For example, in GEinCEE, the CoP was seen as important in providing a space where CEE issues could be discussed and members 'voices' could be heard this was seen in contrast to more heterogeneous spaces: 'in the discussions on gender equality which, as expressed during meetings and workshops, to a great extent are dominated by the Western perspective. This is often underpinned by an assumption that CEE institutions need knowledge transfer and solutions from more developed North-West countries, or the imposition of specific goals of gender equality policies and aspects which they should cover...'(Sekula, Ciaputa, Warat, Krzaklewska, Beranek & Reidl, p. 91 in this volume).

The experiences of GEinCEE together with Alt+G constitute important contributions to a post-colonial perspective on furthering gender equality throughout Europe and beyond. Recent developments such as the political backlash and opposition against gender equality and/or feminist movements occurring across many EU Member States irrespective of their equality trajectory questions the simple transfer model of knowledge and policy from West to East. What is needed rather, are opportunities for exchange and mutual learning in response to shared needs and features of the regional/ national context. Nowhere was this more apparent than in the Alt+G CoP which had been established with the explicit goal to create 'alternative infrastructures' for gender equality in Slovenia to address the lack of official support and resources. Initial activities of this CoP were targeting the national level standards used to evaluate academic excellence, and norms for career promotion. Mihajlović Trbovc in Chapter 6 outlined how CoP members engaged in activities that aimed to improve these national level regulations with gender fair outcomes.

The geographic focus, therefore, emphasises a point already raised by Brown & Duguid (1991) in the introduction of this volume: the ability to anchor the creation of new knowledge and practice in a particular (geographic) context makes CoPs highly effective, organic instruments of innovation as practical solutions are generated where they emerge and are most needed. What the geographic experience, furthermore, has shown is the strength of CoPs to push forward the agenda of gender equality even when governmental support is lacking or the wider political context unfavourable.

Thematic Communities of Practice

Two thematic CoPs were set-up during the project, one focusing on gender budgeting (GenBUDGET) and the other focusing on sustainability (STRATEGIES). Although member organisations in both CoPs clustered in certain EU regions – Northern Europe for GenBUDGET and Continental Europe for STRATEGIES with notable exceptions – the geographical and political context was not a decisive element in the sense of community and shared interest. Rather, specific thematic issues were identified: for the GenBUDGET CoP it meant tackling gender bias in decision making processes by the means of gender budgeting. Despite the fact that CoP members joined the CoP with differing levels of experience and knowledge of gender budgeting, the shared interest in taking this gender equality strategy further in their respective organisations established a common ground for a fruitful collaboration. The shared, common domain for the other thematic CoP, STRATEGIES, focused on gender equality from a sustainability perspective. The collective intention of this CoP was to identify, share, and leverage strategies for sustainable gender equality. Continuities and discontinuities in gender equality projects – formed the basis for the domain of this CoP.

Disciplinary-based Communities of Practice

As regards the disciplinary based CoPs – varying views existed on the extent of the usefulness of the disciplinary based focus. For example, as regarding the LifeSciCoP, in Chapter 7 by Reiland, Palmén and Kamlade, one of the CoP members stated, 'we do not discuss sciences actually. We discuss the policies of the institutes, and my institute is an institute of life sciences but the problems are the same' (p. 124). Other LifeSciCoP members, however extolled the usefulness of the disciplinary approach, particularly in relation to how the representation of women/men differs according to discipline. Other issues, such as the evaluation of excellence could be characterised in a similar fashion: whilst the evaluation of researchers' achievements forms part of the overall science system – it impacts differently according to discipline. For example, in the LifeSciCoP the issue of gender equality and career advancement was perceived as having a disproportionate impact on researchers compared to other disciplines since the Life Sciences are experienced as particularly competitive. The assessment of researchers therefore figured as a priority on this CoPs agenda, something that was less important for GENERA. For the physics-oriented CoP, the disciplinary focus was also cited as the main reason to join the CoP. While physics was appreciated as one of the 'toughest nuts to crack' for gender equality work referring to gender imbalance in representation – the integration of the gender dimension into teaching and research content which was also high on this CoP's agenda (Reiland, Palmén & Kamlade, p. 125 in this volume).

Overall, it is not surprising that a disciplinary approach provides a strong common ground for CoP members to define their 'domain'. The scientific disciplines, characterised each by their unique traditions, practices, procedures, or epistemological preferences, retain much of their power despite the abundant calls for inter- or transdisciplinary work (Kreber, 2010; Krishnan, 2009). CoPs offer in this context the arena for spelling out disciplinary specific challenges and potential solutions to the overarching gender equality issues such as career advancement or the integration of the gender dimension.

Community

The quality of the social relations that characterised the ACT CoPs were seen to operate as a powerful force that was able to harness the push for change. As stated by Mihajlović Trbovc (p. 113 this volume), 'the CoP structure and sense of community provided the framework that turned unforeseen challenges into windows of opportunity for institutional change, and created space for mutual learning'. This support was seen to go on to motivate CoP members to engage with change in their own institutions, 'the sense of community and belonging, and the empathic understanding of the complexity of endeavours to strengthen gender equality gives strength and motivation to continue striving for better work and life conditions for everyone at Vilnius university' (Sekula, Ciaputa, Warat, Krzaklewska, Beranek & Reidl, p. 88 this volume). Key to this sense of community was the recognition that the CoP was seen as a safe space. Axelsdóttir, Steinþórsdóttir & Einarsdóttir (p. 142 this volume) explain how CoP members 'experience GenBUDGET as a social support, wherein they are not alone in their struggle for gender equality and they find it important to stay in contact, and think of the CoP as a way to energise'. This was also echoed and highlighted as being a main benefit of belonging to the GENERA CoP – due to possible changes in university leadership it was recognised that 'if you work on this inside the physics department or the science faculty you become very vulnerable ... suddenly ... you have no support or are very isolated, so a lot of the discussion is ... pure therapy' (Reiland, Palmén & Kamlade, p. 127 in this volume).

That CoPs provide a safe space and sense of community resonated across all CoPs but was particularly important in those where gender equality activities encountered resistance. For example, in describing the experience of the the Alt+G CoP, Mihajlović Trbovc (p. 113 this volume) highlights how the more flexible community approach enabled 'the unstructured sharing of experience among CoP members on problems and challenges they faced in everyday work helped them reflect on the process and functioned as a "moral" support when facing resistances'. Therefore, CoPs not only served as a safe space where moral support was fostered, but also secured a forum to share different strategies against resistance. This was described in the contribution by Sekula, Ciaputa, Warat, Krzaklewska, Beranek & Reidl (p. 94 this volume): 'The CoP's meetings provided opportunities to both share good practices in dealing with resistance, already tested by some partners, and receiving emotional support by those who face reluctance or hostility towards gender equality interventions [...]'.

Another important insight regarding the community aspect of CoPs concerns the need to include men in the work of gender equality. As explored in the introduction to this volume and Chapter 2 by Thomson, Barnard, Hassan & Dainty – most of the ACT CoP members were women. In part, this is due to the direct experience of gender inequality by women, and hence the greater awareness of the gendered nature of academic institutions in which they work (Acker, 2000). Enhancing participation of men was highlighted as major challenge for the CoP by members from the GENERA CoP as women are traditionally under-represented in physics: 'if you are really serious you have got to make sure that you have men who are getting involved as well, and that they are shouldering the work...I think it shows that it is being taken seriously at the discipline level and it sends a signal'

(Reiland, Palmén & Kamlade, p. 127 in this volume). The further inclusion of men would be key in order to achieve a more just distribution of gender equality work especially in the light of recent developments such as the introduction of the GEP eligibility criterion to access Horizon Europe funding (European Commission, 2021). The advantage of having a GEP and, therefore, being able to access Horizon Europe funding benefits both men and women alike while most equality work is still predominantly carried out by women (Tzanakou & Pearce, 2019).

A central insight documented in Chapter 5 by the GEinCEE CoP speaks about not only the internal sense of community among CoP members but also the increased engagement with the extended, wider organisational community. Sekula, Ciaputa, Warat, Krzaklewska, Beranek & Reidl highlight how CoP members can be isolated in their own institutions while their CoP membership can help forge strategic alliances with staff and students. The STRATEGIES CoP intentionally included networks as part of its CoP (Eurodoc, CPED, Femmes & Sciences, Femmes et Mathématiques). Damala, Mour & Godfroy emphasise in Chapter 9 how this has proved to be a particularly useful approach to disseminate news, surveys, best practices and guidelines for researchers and practitioners.

As well as extending allies and making new contacts, extending the community to leadership and decision-making levels was recognised as key. Chapter 5 by Sekula, Ciaputa, Warat, Krzaklewska, Beranek & Reidl describe how in only one GEinCEE member's institution, representatives from the managerial board and director's circle became interested and engaged in the gender equality work, while in most other member institutions this proved extremely difficult. The difficulty to garner leadership support poses one of the central questions alluded to in the introduction regarding the wider organisational embedding of CoPs: beyond their potential to provide a safe space among participants, what is the potential of CoPs to affect the organisational structures and power relations which perpetuate gender inequality. There is little doubt that CoPs can provide alternative solutions by building alliances 'on the ground' among gender equality practitioners, researchers, administrative staff, and others when national infrastructures are absent. The quality of the community was identified by CoP members as especially important and valuable for this task. However, especially for CoPs working towards greater gender equality, the collaboration and mutual learning, clearly has a political dimension that needs to be tied to political activism and a push for change, as argued by Thomson, Barnard, Hassan, Dainty in Chapter 2.

Practice

The practice dimension has formed an integral part of each CoPs work. As highlighted in the introduction, a practice or a 'shared repertoire' can include 'routines, words, tools, ways of doing things, stories, gestures, symbols, genes, actions or concepts' (Wenger, 1998, p. 83) that materialise past actions yet can be built on for current and future activities. Reidl, Baranek & Holzinger, in Chapter 10 highlight how this 'practice' is based on a learning process – developed through interaction and participation – whereby knowledge is constructed on the basis of action. They also go on to note the synergies of the CoP approach with gender equality and institutional change where the latter requires precisely the remodelling of established practices (Bruni et al., 2004; Gherardi et al., 2018). The chapter by Mihajlović Trbovc provides a prime example along these lines, as it describes how concerns within the CoP shifted from tackling national level regulations to everyday practices regarding gender sensitive language in institutional level documents and websites. Changing these practices is not an easy task as they constitute the very processes and mechanisms that are re-inscribed in daily operations of the organisation. What's more, this not only concerns the daily formal operations but also the often unconscious habits and unspoken rules underlying these organisational procedures. As Angela Wroblewski & Rachel Palmén (2022) underscore: despite the enactment of changes in formal procedures to counteract bias in recruitment – outcomes may not be affected if practices remain unchanged. The insistence on the importance of gender equality *practitioners* acquires renewed significance in this context: as change agents to modify and provide alternative formal and informal organisational procedures but above all the need to engage in, and catalyse disruptive practices for real change.

On one level, the exchange of good institutional practices became a key focus of the CoPs. In this sense, 'good practice' was conceptualised distinctly as something more formal and institutionalised compared to the 'practice' dimension of a CoP. Nevertheless, the CoPs' focus on exchanging and sharing good practice formed an important element of their work portfolio. It was recognised that the scope for this exchange can vary according to the scope of the CoP. For example, CoPs that span Europe cited the advantages of sharing good practices across Member States where in some countries advanced legislation and policy frameworks for gender equality can inspire action in countries where these are lacking. In the LifeScience CoP, a concrete example of how sexual harassment protocol was shared from an institution that had a very well developed legislative and policy framework to an institution operating in a context without this developed framework was recognised as enabling the transfer of good practice (see Chapter 7 by Reiland, Palmén, Kamlade). The sharing of good practices was accepted to be of substantial importance in order to preserve efforts and energy without reinventing the wheel.

The 'Targeted Implementation Projects' (TIPs) promoted and implemented by the GenBUDGET CoP provide a second interesting approach to gender equality practice. TIPs precisely zoom in on the organisational mechanisms by which institutional budget is formally and informally allocated. Axelsdóttir, Steinþórsdóttir & Einarsdóttir in Chapter 8 described how the TIPs have provided the structure for each CoP member to really push for institutional change and modify practices within their institution. Different TIPs were established by CoP members focusing for example on the allocation of financial funds, the gender pay gap, workload allocation schemes, internal research grant processes, the status of seasonal teachers and the status of PhD graduates. At the University of Birmingham (UK), the implemented TIP showed for example that the 'workload allocation system was presented as a tool to enable transparency and fairness' while in reality it benefits men. A similar insight resulted from the TIP implementation at the Carlos III University in Madrid (Spain) which showed gendered outcomes of pay mechanisms. The TIPs of GenBUDGET thus provide a unique window on advantages as well as challenges to identify organisational practices that have clearly gendered implications. By pooling their experiences, the CoP members in this case not only identified those practices but also harnessed their collective experience and creativity to suggest alternative, more gender equal administrative procedures.

Finally, the development and implementation of the Gender Equality Audit and Monitoring (GEAM) tool provides a third example how CoPs can give rise to a shared practice that advances gender equality. The GEAM, as described in Chapter 3 provides a standardised environment for carrying out survey-based gender equality audits in academic organisations. It has been developed, tested and above all used so far by 22 member organisations from different CoPs. The fact that it is a standardised tool provides a common foundation for CoP members to engage in a discussion about its' application and customisation, the interpretation of its results as well as its weakness and strength. The shared tool constitutes a 'boundary object' (Star, 1989) that binds together different practitioners across organisational – but also national boundaries in their quest to insert the GEAM in their institutional GEP practices. Practice involves here the technical aspects of setting up and customising a GEAM survey but it also refers to the knowledge to generate reports, interpret the findings and convert statistical results into a custom gender equality strategy and action plan. Since the GEAM can be used during the initial GEP audit but also for long-term monitoring of staff experiences of gender (in)equality, it is likely to become a building block of the GEP data monitoring infrastructure within organisations. The CoP level thereby functions as a shared space to become familiar with the tool as well as generating new knowledge as when CoP member institutions analyse their respective results in a comparative fashion. The example of the GEAM, therefore, can be seen as a good illustration of Silvia Gherardis' point alluded to in the introduction (Gherardi, 2009), that it is the practice that constitutes the community.

Communities of Practice for greater gender equality, its strength, and challenges

This book builds on two distinct bodies of literature, the CoP literature which spans many disciplines (business, management, higher education, community development, etc.) as well as the literature looking at implementing gender equality interventions in R&I and HE. Initial readings of both bodies of literature showed grounds for hope. Despite a general lack of crossover, synergies between the two bodies of literature indicated that a CoP approach may be a useful mechanism to advance gender equality and institutional change. The literature considering the effectiveness of gender equality interventions in R&I identified those success/hindering factors that are pivotal to implementation processes. Factors like governance framework, top-management commitment, bottom-up participation, framing synergies with other initiatives, resistance, resources, sustainability, gender competence, transparency, targets and monitoring as well as accessible data and information – have all been identified as impacting on implementation processes (Palmén & Kalpazidou Schmidt, 2019). A cursory reading of the CoP literature with its emphasis on practice, competence development, mutual learning as well as community engagement, participation, sharing, consensus (Cambridge & Suter, 2005) suggests that a CoP approach might be useful to advance the gender equality agenda (Thomson et al., 2021). We have been fortunate enough throughout the course of the ACT project – to be able to test this assumption and collect and document the relevant evidence as to the extent to which a CoP approach can be useful for advancing gender equality efforts in R&I and HE.

One of the central tensions in the ACT project highlighted in the introduction – aimed at strengthening gender equality within R&I and HE organisations but through using an inter-organisational CoP approach. The majority of the CoP literature (specifically coming from business, management, higher education) charts CoPs within one company, organisation, or institution. We however, set up and facilitated inter-organisational CoPs yet with the overall objective of stimulating and sustaining change within an institution. As we have seen throughout some of the chapters in this book, the power of inter-organisational CoPs can be harnessed for institutional change. In the context of GEinCEE – Sekula, Ciaputa, Warat, Krzaklewska, Beranek & Reidl describe how they have managed to reconcile this tension on the ground by specifically emphasising the regional focus of their CoP. Setup as an inter-organisational collaboration, the CoP provides 'intermediary support structures' where diverse initiatives can cross-fertilize and build up capacity for structural change through local change agents. As the authors go on to clarify, 'such structures may as well allow for the exchange of localised, context-specific knowledge and discuss tailored strategies that are possible in the region or national context'. (p. 83 in this volume). This approach is echoed by Mihajlović Trbovc in her chapter charting the experience of the Alt+G CoP operating in Slovenia where she describes how the inter-organisational CoP enabled the 'fast transfer of knowledge and practices' within the Slovenian academic community. As a result of this exchange of practices "the quality of institutional changes that took place in individual organisations' (p. 112 in this volume) was enhanced.

At the same time, CoPs are best seen as an additional support mechanism to enhance institutional efforts, i.e., institutions need both financial and human resources to successfully engage with institutional change. The tranche of funding provided by the European Commission for structural change projects has to date been a remarkably welcome source, especially in countries where gender equality in R&I is neglected. As commented by Mihajlović Trbovc (p. 113 this volume): 'since the CoP approach operates on the fuel of personal motivation and depends on individual rather than institutional commitment, its ability and reach in enhancing institutional change is contingent on favourable structural context'.

It also rather doubtful to what degree CoPs can provide an effective means to counter the strong backlash against gender equality in many Member States. Although CoPs provide a safe space for its members and help to overcome the isolation experienced by many gender equality practitioners, this might not be enough when confronted with outright hate. As reported in Chapter 5, 'the context of anti-gender discourse and initiatives prominent in some of the CEE countries was perceived by the CoP members as having a negative impact on their work and the possibility to implement gender equality interventions' (p. 94 this volume). The hostile climate not only affected the motivation of certain CoP members but also resulted in self-censorship regarding the naming of their activities, avoiding for example 'gender' and replacing it with less contested concepts such as 'equality between women and men'. The experience throws new light on the issue of power relations and CoPs – not so much among its members but rather regarding the embedding of the CoP within their organisation and the wider political environment.

What cannot be emphasised too much when considering CoPs as a means to achieve greater gender equality is the importance of adequate resourcing. Given the spirit of bottom-up organising in combination with a strong intrinsic motivation, CoPs are easily misconceived as requiring little to no formal management. In stark contrast, the experience of the ACT CoPs and especially their facilitators showed that establishing a sense of community, achievement, and vitality among the members requires a lot of energy, time, and commitment. Our experience mirrors the recent insights published in the *Communities of Practice Playbook* by the EC, whose authors remark that community management is far from an effortless affair. Rather, 'community management is often perceived as cumbersome and lacking in resources (time/recognition)' (Catana et al., 2021, p. 14). A community cannot work without systematic management that involves practical issues such as convening meetings or more substantive issues such as knowledge synthesis and brokerage among others. The importance of providing adequate resources either through direct sponsoring schemes or earmarking time for CoP facilitation is especially important in the context of gender equality work. First, because gender equality work has been historically and in general under-financed, obliging many women to drive gender equality agendas on a voluntary basis. Secondly, because the effort necessary for cultivating a community is easily rendered invisible as it constitutes another case of 'affective labour' where the nurturing of caring relations is seen as genuine feminine qualities for which neither knowledge nor effort and hence resources seem necessary (Daniels, 1987). Since relational, affective, and interpersonal aspects not just within CoPs but also within the wider academic organisations are rarely explicitly acknowledged in budgetary planning or career/reward systems, there is a real danger that engagement in CoPs becomes another item on the invisible service work carried out predominantly by women (Guarino & Borden, 2017; Social Sciences Feminist Network Research Interest Group, 2017).

CoPs offer an alternative space within contemporary academic organisations. By their very definition, CoPs require a certain degree of autonomy when establishing their 'domain' in response to actual needs and interests of their members. As such, they do not fully overlap with official organisational strategies, procedures, incentives, bureaucratic requirements, or established performance targets. CoPs, at least by their very idea, present a line of flight beyond the academic work environment with its precarious yet highly demanding and performance-oriented productivity regime. This split-second of autonomy – to work freely and directly in response to concrete (gender equality) needs without having to worry about paper work, reporting, impact factors, meeting of targets – is motivating and rewarding in itself. Indeed, the very setup of the ACT project allowed CoPs to concentrate on their agenda without being over-burdened by deliverables and reporting tasks. They offer the opportunity to go back to 'doing the doing' and engage in actual equality work rather than 'doing the document' as a box-ticking exercise (Ahmed, 2007). And yet, in the same way that feminist scholars and gender equality practitioners have fought for the adequate recognition of 'service' work within academia, CoP work needs to be part of this picture – adequately resourced and recognised. After all, to the degree that CoPs are part of a struggle for greater gender equality, they form part of a struggle for equal distribution of resources and recognition (Fraser, 2003).

References

Acker, Joan (2000). Gendered Contradictions in Organizational Equity Projects. *Organization*, 7(4), 625–632.

Ahmed, Sara (2007). 'You End Up Doing the Document Rather Than Doing the Doing': Diversity, Race Equality and the Politics of Documentation. *Ethnic and Racial Studies*, 30(4), 590–609. https://doi.org/10.1080/01419870701356015

Brown, John Seely, and Duguid, Paul (1991). Organizational Learning and Communities-of-Practice: Toward a Unified View of Working, Learning, and Innovation. *Organization Science*, 2(1), 40–57. https://doi.org/10.1287/orsc.2.1.40

Bruni, Attile, Gherardi, Silvia, & Poggio, Barbara (2004). *Gender and Entrepreneurship: An Ethnographic Approach*. Routledge. https://doi.org/10.4324/9780203698891

Cambridge, Darren, & Suter, Vick (2005). *Community of Practice Design Guide: A Step-by-Step Guide for Designing & Cultivating Communities of Practice in Higher Education*. EDUCAUSE Learning Initiative (ELI) & Virtual Learning Environment (VLE). https://bit.ly/3lF3LMc

Catana, Carolina, Debremaeker, Iris, Szkola, Susanne, & Williquet, Fredric (2021). *The Communities of Practice Playbook: A Playbook to Collectively Run and Develop Communities of Practice*. Publications Office. https://data.europa.eu/doi/10.2760/42416

Daniels, Arlene Kaplan (1987). Invisible Work. *Social Problems, 34*(5), 403–415.

European Commission. (2021). *Horizon Europe Guidance on Gender Equality Plans*. Publications Office of the European Union. https://data.europa.eu/doi/10.2777/876509

Fraser, Nancy (2003). *Redistribution or Recognition? A Political-Philosophical Exchange*. Verso.

Gherardi, Silvia (2009). Community of Practice or Practices of a Community. In S. J. Armstrong & C. V. Fukami (Eds.), *The Sage Handbook of Management Learning, Education, and Development* (pp. 514–530). SAGE Publications Ltd. https://books.google.es/books?id=Om3nZSDGKNUC

Gherardi, Silvia, Cozza, Michela, & Poggio, Barbara (2018). Organizational Members as Storywriters: On Organizing Practices of Reflexivity. *The Learning Organization, 25*(1), 51–62. https://doi.org/10.1108/TLO-08-2017-0080

Guarino, Cassandra M., & Borden, Victor M. H. (2017). Faculty Service Loads and Gender: Are Women Taking Care of the Academic Family? *Research in Higher Education*, 1–23. https://doi.org/10.1007/s11162-017-9454-2

Kreber, Carolin (2010). *The University and its Disciplines: Teaching and Learning Within and beyond Disciplinary Boundaries*. Routledge.

Krishnan, Armin (2009). *What are Academic Disciplines? Some Observations on the Disciplinarity vs. Interdisciplinary Debate* (No. 03/09; NCRM Working Paper Series). ESRC National Centre for Research Methods. https://ams-forschungsnetzwerk.at/downloadpub/what_are_academic_disciplines2009.pdf

Palmén, Rachel, & Kalpazidou Schmidt, Evantia (2019). Analysing facilitating and hindering factors for implementing gender equality interventions in R&I: Structures and processes. *Evaluation and Program Planning, 77*, 101726. https://doi.org/10.1016/j.evalprogplan.2019.101726

Social Sciences Feminist Network Research Interest Group. (2017). The Burden of Invisible Work in Academia: Social Inequalities and Time Use in Five University Departments. *Humboldt Journal of Social Relations, 39*, 228–245.

Star, Susan Leigh (1989). The Structure of Ill-Structured Solutions: Boundary Objects and Heterogeneous Distributed Problem Solving. In L. Gasser & M. N. Huhns (Eds.), *Distributed Artificial Intelligence* (pp. 37–54). Morgan Kaufmann. https://doi.org/10.1016/B978-1-55860-092-8.50006-X

Thomson, Aleksandra, Palmén, Rachel, Reidl, Sybille, Barnard, Sarah, Beranek, Sarah, Dainty, Andraw R. J., & Hassan, Tarek M. (2021). Fostering Collaborative Approaches to Gender Equality Interventions in Higher Education and Research: The Case of Transnational and Multi-Institutional Communities of Practice. *Journal of Gender Studies, 0*(0), 1–19. https://doi.org/10.1080/09589236.2021.1935804

Tzanakou, Charikleia, & Pearce, Ruth (2019). Moderate Feminism within or against the Neoliberal University? The Example of Athena SWAN. *Gender, Work & Organization, 26*(8), 1191–1211. https://doi.org/10.1111/gwao.12336

Wenger, Etienne (1998). *Communities of Practice: Learning, Meaning, and Identity.* Cambridge University Press. https://books.google.es/books?id=heBZpgYUKdAC

Wenger, Etienne, McDermott, R. Arnold, & Snyder, William (2002). *Cultivating Communities of Practice: A Guide to Managing Knowledge.* Harvard Business Press. https://books.google.es/books?id=m1xZuNq9RygC

Wroblewski, Anglea, & Palmén, Rachel (2022). A Reflexive Approach to Structural Change. In A. Wroblewski & R. Palmén (Eds.), *Overcoming the Challenge of Structural Change in Research Organisations: A Reflexive Approach to Gender Equality.* Emerald Group Publishing.

Index

Note: Page numbers in **Bold** refer to tables; and page numbers in *italics* refer to figures

For Product Safety Concerns and Information please contact our EU
representative GPSR@taylorandfrancis.com
Taylor & Francis Verlag GmbH, Kaufingerstraße 24, 80331 München, Germany